T0321197

INTERFEROMETRY IN RADIOASTRONOMY AND RADAR TECHNIQUES

Interferometry in Radioastronomy and Radar Techniques

by

R. WOHLLEBEN

H. MATTES

and

TH. KRICHBAUM

Max-Planck-Institut für Radioastronomie,
Bonn, F.R.G.

KLUWER ACADEMIC PUBLISHERS

DORDRECHT / BOSTON / LONDON

Library of Congress Cataloging-in-Publication Data

Wohlleben, Rudolf, 1936-
    [Interferometrie in Radioastronomie und Radartechnik. English]
    Interferometry in radioastronomy and radar techniques / Rudol[l]f
Wohlleben, Hartmut Mattes, Thomas Krichbaum.
        p.   cm.
    Translation of: Interferometrie in Radioastronomie und
Radartechnik.
    Includes bibliographical references (p.   ) and index.
    ISBN 0-7923-0464-0 (HB : acid free paper)
    1. Radio interferometers. 2. Radio astronomy. 3. Interferometry.
I. Mattes, Hartmut, 1942-   . II. Krichbaum, Thomas, 1959-   .
III. Title. IV. Title: Interferometry in radio astronomy and radar
techniques.
QB479.3.W64 1990
522'.682--dc20   90-5309

ISBN 0-7923-0464-0

Published by Kluwer Academic Publishers,
P.O. Box 17, 3300 AA Dordrecht, The Netherlands.

Kluwer Academic Publishers incorporates
the publishing programmes of
D. Reidel, Martinus Nijhoff, Dr W. Junk and MTP Press.

Sold and distributed in the U.S.A. and Canada
by Kluwer Academic Publishers,
101 Philip Drive, Norwell, MA 02061, U.S.A.

In all other countries, sold and distributed
by Kluwer Academic Publishers Group,
P.O. Box 322, 3300 AH Dordrecht, The Netherlands.

Translators: RUDOLF WOHLLEBEN, DAVID A. GRAHAM

Printed on acid-free paper

Printed in the Netherlands

TABLE OF CONTENTS

Preface   xi

Foreword   xiii

List of Symbols   1
    Latin Symbols   1
    Greek Symbols   6

1. Introduction   8

2. Characteristic magnitudes of cosmic radio radiation   11

3. Reception of quasi-monochromatic, partially polarized
    plane waves   13

4. Reception of randomly polarized waves from an
    extended, incoherent source   16

5. Theory of linear spatial filters   18

6. Interferometric techniques   25
    6.1. Principle of the interferometer   25
    6.2. Effect of bandwidth   28
    6.3. Effect of extension of the radiations source on
          the power pattern   30

7. Radioastronomical interferometers   34
    7.1. Sea interferometer   34
    7.2. Interferometer with passive reflectors   34
    7.3. Ryle's twin interferometer   35
    7.4. Phase-switched interferometer   35
    7.5. Swept-frequency interferometer   36
    7.6. Lobe sweeping interferometer   38
    7.7. Multi-element or grating interferometer   38
    7.8. Christiansen-cross and Mills-cross interferometer   41
    7.9. Correlation systems   44
          7.9.1. Hanbury Brown-Twiss or intensity interferometer
                  (post detection correlation interferometer)   51
          7.9.2. Compound intensity interferometer   53
          7.9.3. Very long baseline interferometer and aperture synthesis 56
                  7.9.3.1. Introduction (VLBI)   56
                  7.9.3.2. Applications of VLBI   59
                  7.9.3.3. Interferometer fundamentals   61
                  7.9.3.4. Coherence theory   63

7.9.3.5.   The response of a two-element crosscorrelation
           interferometer   66
7.9.3.5.1.   The fringe washing function   68
7.9.3.6.   The interferometer geometry   70
7.9.3.6.1.   The interferometer response to an
             extended source   72
7.9.3.6.2.   The $u - v$-plot   75
7.9.3.7.   Single and double sideband systems   79
7.9.3.7.1.   The single sideband system (SSB)   80
7.9.3.7.2.   The double sideband system (DSB)   82
7.9.3.8.   A working VLBI system   83
7.9.3.8.1.   Frequency standards   84
7.9.3.8.2.   Time synchronization   84
7.9.3.8.3.   Data recording   87
7.9.3.8.4.   The correlator center   88
7.9.3.9.   Aperture synthesis mapping   88
7.9.3.9.1.   Introduction   88
7.9.3.9.2.   Calibration   88
7.9.3.9.2.1   The closure phase   89
7.9.3.9.2.2. The closure amplitude   90
7.9.3.9.2.3. Selfcalibration   91
7.9.3.9.3.   The Fourier inversion of the
             visibility function   91
7.9.3.9.3.1. The weighted sampling function   96
7.9.3.9.3.2. Gridding   98
7.9.3.9.3.3. The clean algorithm   99
7.9.3.9.3.4. The maximum entropy method
              (MEM)   100
7.9.3.10. Further applications of VLBI   101
7.9.3.10.1. Spectral line VLBI   101
7.9.3.10.1.1. Principle   101
7.9.3.10.1.2. The phase of the signal   104
7.9.3.10.1.3. Phase referencing   105
7.9.3.10.1.4. The fringe rate method   105
7.9.3.10.2. The application of VLBI for
             astrometry and geodesy   106
7.9.3.10.2.1. Introduction   106
7.9.3.10.2.2. Sensitivities to delay and
               fringe rate   111
7.9.3.10.2.3. Calculation of source position
               (and) or baseline coordinates   112
7.9.3.10.2.4. The retarded baseline   114
7.9.3.10.3. Space VLBI   115
7.10.   Signal processing in a phase coherent interferometer   118
7.10.1. Signal processing at a single frequency   119
7.10.2. Single-sideband (SSB) single processing in a
         finite bandwidth   122
7.10.3. Consequences of delay tracking in the SSB finite
         bandwidth case   123
7.10.3.1. Double sideband processing   123

	7.10.4.	Possible ways of separating the sidebands in case of DSB reception	125
		7.10.4.1. Sideband separation by phasing	125
		7.10.4.2. Sideband separation by fringe rate	126
		7.10.4.3. Concluding remark on sideband separation	126

8. Radar applications of interferometry — 130
8.1. Introduction — 130
8.2. Fundamentals — 130
   8.2.1. Doppler shift — 131
   8.2.2. Defocussing effects generated by target movement — 132
   8.2.3. Resolution of the angular ambiguity — 135
   8.2.4. Resolving power and ambiguity function — 143
      8.2.4.1. Radar receiver optimization — 143
      8.2.4.2. Derivation of the ambiguity function (af) — 148
      8.2.4.3. Pulse compression — 154
      8.2.4.4. Signal form and antenna pattern — 157
      8.2.4.5. Ambiguity function of radar systems with a linear antenna array or linear interferometer — 160
         8.2.4.5.1. Signal bandwidth and ambiguity — 160
         8.2.4.5.2. Transfer function — 165
         8.2.4.5.3. The ambiguity function for simultaneous operation — 167
   8.2.5. Influence of the atmospherical refraction on the angular measurement — 169
8.3. Two-dimensional phase-comparison monopulse radar — 175
8.4. Phase-comparison monopulse radar for three-dimensional target tracking — 176
8.5. Pseudo-noise code interferometer — 177
8.6. Interferometer systems for synchronous satellite navigation — 181

9. References — 189

10. Index — 202

Photograph 1

World's largest fully steerable radio telescope (built 1972) in Bad Muenstereifel-Effelsberg
(Eifel Mountains, near Bonn)
Diameter: 100.m
Total Height: 94.m
Total surface accuracy (rms) of the reflector: 0.5.mm (1986)
Modes of operation:
1. Primary focus: $f_p/D_p = 0.3$
2. Gregorian secondary focus: $f_e/D_p = 3.87$
Magnification: 12.9
Track ring diameter: 64.m
Azimuth movement range: ± 360°
Elevation movement range: + 7. −+ 94.°
Total weight: 3200.t
Pointing accuracy: 2." (night), 20." (day)
Max. slewing speed (azimuth, elevation): 20°/min.
Highest frequency: 43.GHz (1985)
Operated by: Max-Planck-Institut für Radioastronomie, D-53 Bonn 1, F.R.Germany

Photograph 2

Short Baseline Interferometer (SBI) of 12 rotational paraboloidal reflector elements for supersynthesis (aperture synthesis by earth rotation) observations in Westerbork (Netherlands)

Total baseline length: 2700.m
Element diameter: 25.m
Reflector surface: wire mesh of 8.mm distance and 0.8.mm diameter
Highest frequency: 5.GHz
Pointing accuracy: $\pm 30,$"
Rotational velocity: 0.125 to 0.50°/min
Mode of operation: primary focus
Orientation of the rail tracks: East-West
Number of fixed elements: 10 (B" = 144.m), 2 elements movable (± 150.m)
Operated by: Dutch foundation for radioastronomy (Z.W.O.)

PREFACE

The present book: Interferometry in radioastronomy and radar techniques, covers the various applications of interferometry which followed the pioneering experiment of Michelson in 1922. The first radioastronomical measurements, beginning with the short-wave observations of Karl Jansky in 1930/31 were made with single antennas, but the need for increased directivity and sensitivity led to the use of interferometric methods in this field, notably by Martin Ryle and co-workers in the U.K. and Peter Wild, W.N. Christiansen and B.Y. Mill in Australia. The technique developed quickly and soon led to antenna systems spanning the globe for VLBI measurements.

In the year following the publication of the first German edition of this book, Sir Martin Ryle/Cambridge received the Nobel Prize in Physics for the technique of aperture synthesis, from his first ideas in 1952/53 to the construction of the one-mile telescope in 1960 and the five-mile telescope in 1972.

Ryle himself, speaking on the occasion of the presentation of the Nobel Prize, stated that some of the groundwork was laid by M. Blythe in 1946.

Meanwhile the Very Large Array (VLA) near Socorro, N.M. has come in operation and many conferences (1980) on interferometry have taken place. Worldwide hundreds of radioastronomers are working with interferometry, so that the time seems ripe for the appearance of the present book by R.W., H.M., and T.K. The text will be a helpful tool for beginners at the graduate student level.

Richard Wielebinski

(Executive Director
Max-Planck-Institut
für Radioastronomie
Bonn, F.R.G.)

FOREWORD

Since the invention of the optical telescope in the 17th century attempts have been made to improve the resolution of such instruments in order to resolve double stars and clusters. The interferometer represents such an attempt.

The first radioastronomical measurements were made in 1931, and applications of interferometry came soon afterwards in 1947. The rise of space science in subsequent years led to a fruitful cooperation between radioastronomy and radar leading to improved resolution in position and range.

These facts and the obvious need for a textbook for graduate students in this new field were the decisive arguments for the 1973 German edition. Requests by visitors, observers and interested amateurs for an English edition led us to produce the present text, revising the whole book and completely rewriting the sections on aperture synthesis and very long baseline interferometry (VLBI) to reflect recent developments; this present work is therefore a revised edition of the German edition, which is still available from the authors.

In recent years radioastronomical aperture synthesis and VLBI have become a very powerful tool. These methods meanwhile have grown to great complexity and it was our goal to give a comprehensive presentation within a few chapters. It is beyond the scope of this book to cover all applications and data reduction algorithms of aperture synthesis and VLBI. The reader interested in further details may refer to another textbook (292).

This text is intended for radio astronomers and radar engineers but can also serve as an introduction to the subject for more generally interested scientists. Physicists with some experience of optical interferometry will notice that the present book does not contain a discussion of the wave/particle duality of light, of the methods for the measurement of the velocity of light and other topics which can be summarized under the heading "laboratory interferometry". Here exhaustive presentations by Steel (43) or (22,45,69) are available, so that we exclude such topics.

In reading the present monograph some experience in complex variables, statistical averaging and in signal theory (for the radar part, chapter 8) will be helpful.

The authors acknowledge critical comments on the first edition by colleagues at our Max-Planck-Institut für Radioastronomie/Bonn, especially to Dr. Wolfgang Reich, Dipl.-Phys. Christian A. Hummel, Dr. David A. Graham and by Dr. Karl-Heinz Gebler (Institute for Radioastronomy, University of Bonn) concerning the radioastronomical parts and Professor Alexander Wasiljeff of the Electrical Engineering Department of the University of Bremen for the radar chapter. These comments are incorporated in the present edition. Professor Richard Wielebinski undertook the task of writing the preface, for which we are very grateful. The authors are very grateful to Dr. David A. Graham, who checked the

English and finally we thank Mrs. Christine Tilly-Schroeder and Ute Runkel for typing
the text of the manuscript.
Bonn, June 1989   Rudof Wohlleben
                                                                                          Hartmut Mattes
                                                                                          Thomas Krichbaum

LIST OF SYMBOLS

Latin symbols

A	numerical aperture, angular resolution, track ellipse constant
$\underline{A}$	complex vector
A_e	effective antenna aperture area/m^2
$A_{\text{EW,NS}}$	resolution in EW,NS direction
A_f	delay constant
$A_{ijkl}(t)$	closure amplitude (for four elements $ijkl$)
$A_{(l,m)}$	source brightness distribution
A_m	maximum value of Ae/m^2
a	radius
a_0	earth radius
a_{ij}	offset term
B	brightness distribution
B_f	delay constant
B'	length of projected baseline
B''	brightness temperature, coordinate of second interferometer element
ΔB	baseline length difference
B_c	area brightness temperature of unpolarized radiation
$B_{k,s}$	brightness distribution (s)
$B_f(l,m)$	source brightness distribution
$B_{m,n}$	complex fringe washing function
$B_{m,n}(\Delta\tau)$	bandpass envelope
B_p	area brightness temperature of polarized radiation
B_t	total baseline length
b	relative bandwidth
$\vec{b}$	difference vector in the source region
$b(f,\tau)$	fringe washing function
C	integral over the square of the autocorrelation function (acf), track ellipse constant (semiaxis)
C_e	refraction parameter
$C_{ijk,\text{obs}}(t)$	sum value, closure phase
C	offset delay constant
$C_{(l,m)}$	source brightness (square absolute value)
$C_{l,m,n}$	phase closure coefficients
$c,\ c_0$	velocity of light (electromagnetic waves) in free space

1

D	reflector diameter; absolute value of distance; baseline vector
$D(l,m)$	set of delta functions
D_k	density weight
D_x	baseline vector component in x-direction
D_λ	baseline distance, normalized on the wavelength
$\vec{d}$	projected baseline vector
d	distance between object points to be separated, small diameter
$E(\vartheta)$	electrical field intensity (vector) in the farfield
E'	energy content of a signal
$E(l,t)$	electrical field, linear-polarized and scalar
ΔE	energy in the frequency interval Δf
$\vec{E}_i$	vector of incident electrical field
E_ϑ	component of E_i
E_φ	component of E_i
$\vec{e}_\vartheta$	unit vector of zenith angle
$\vec{e}_\varphi$	unit vector of azimuth angle
$F(f)$	IF filter function (spectrum), passband
$F(\vartheta)$	fringe pattern of an interferometer
$F(s)$	Fourier transform of $f(t)$
$F_{m,n}(f)$	filter (and amplifier) passband
$f_{A,B}$	carrier frequency of MASER A, B
f	sampling frequency
f_0	center frequency
f_d	Doppler frequency
f_F	frequency shift between two elements
f_{LO}	local oscillator frequency
$f(t)$	fringe rate
f_{RF}	radio frequency/Hz
$f_s = 1/\Delta t$	sampling frequency/Hz
$f\Delta(t)$	signal time function
Δf	bandwidth/Hz, interference fringe distance
Δf_f	fringe frequency offset or resolution
G	directivity (gain)
G_i	gain of i-th antenna (interferometer element)
$G_{i,j}$	complex gains of elements i, j
G_{ij}	baseline gain factor
HA	hour angle
$H(f)$	network transfer function (spectral)
$H(t)$	hour angle of a source
$H_R(\omega, \Delta\xi)$	transfer function of a receiving antenna
$H_T(\omega, \Delta\xi)$	transfer function of a transmitting antenna
h	hight of an element over sea level $/m$

$\vec{h}$	vectorial, effective light
h_b	hour angle of a baseline vector
h_ϑ	component of h
h_φ	component of h
$h(t)$	impulse response
I_e	input current $/A$
$I_{1,2}$	intensity contributions from pinholes $1, 2$ (interferometer)
$I_{\max}$	maximal intensity
IHA	interferometer (baseline) h̲our a̲ngle
$I_{m,n}$	absolute value of the current in the m-th or n-th element $/A$
$I(x)\exp\big(j(x)\big)$	current distribution function with $x = 1\lambda$
$J(l,m)$	source brightness
J_{clean}	clean map
$J_D(l,m)$	dirty image (map)
J_D	dirty image factor
k	Boltzmann's constant, constant, baseline in free space wavelenghts
L	total length of a (linear) aperture (one-dimensional) $/m$
$L(\varphi)$	element pattern $/V/m$
$l_{1,2}$	direction cosinus of (from station 1 to 2)
$\underline{M}$	complex array factor (pattern)
$M(\delta_0, h_0)$	coordinate transform matrix
m	order of interference, source number, variable in the source region
N	n̲oise (power) amplitude, number of array elements, of pixels, of iterations
N'	averaged noise power at a matched receiver output
N_0	spectral noise power density
n	number of (source) transits, element number, index of refraction, number of closure phases, variable in the source region
n_0	index of refraction at the earth surface
$n(s)$	index of refraction depending on s
P	total power density incident on an antenna, power in a synthesis interferometer, farfield point, phase reference point, beam(s)
$P(l,m)$	synthesized beam
$P_{a,b}$	product of element power patterns
$P_{1,2}$	positions of pinholes
$P_{\min.}$	mi̲ni̲mal receiver power $/W$

P_i	reconstructed image (map)
P_{clean}	clean beam
P_t	transmitter power $/W$
p	level of grating lobes, polarization ratio
p_0	amplitude distribution of the n-th array element
$p(x)$	probability density as function of x
Q	probability of false alarm
R	fringe amplitudes, distance between center of aperture and target, far field range (Rayleigh distance)
R_F	limit between near field and far field (of an antenna)
$R(u,v)$	range in $u - v$-plane
R_f	signal-to-noise ratio
R_k	signal-to-noise weight
R_{max}	maximum radar range
R_s	input resistance of an antenna (element)
r	(far field) distance $/m$; radius of two-dimensional Gaussian function
$r(t)$	real part of the acf
$r(x)$	temporal average value of the intensity fluctuation
$r_{12}(\tau)$	(cross)correlation function of $V_1(t)$ and $V_2(t)$
r_{12}	station position vector(s)
r_{LSB}	crosscorrelation function of lower sideband
r_{USB}	crosscorrelation function of upper sideband
S	total power (radiating intensity)
$S(f)$	Fourier spectrum of $s(t)$
$\vec{S}$	unit vector to the (extended) source
S_0	source centrum vector
$S(u,v)$	difference of observed visibility; sampling transfer-function
$S'(u,v)$	weighted sampling function
$\vec{S}$	unit vector (pointing to point P in the sky)
$S_{1,2}$	entropy functions; cross-spectral function
$S_{1,2}(\omega,t)$	crosscorrelation spectrum (ccs)
$\vec{s}$	point source vector
$\vec{s}_0$	unit length vector (element-source)
$\vec{s}_{a,b}$	vector to MASER A, B
$\sin c(x)$	$= (\sin x)/x$ function
s	spatial frequency, index for "statical"
s_k	limiting spatial frequency
$s(t)$	real time signal
$s_0(t)$	real output signal
T	mutual coherence function (mcf), registration time $/s$
T_A	antenna temperature $/K$
T_{Ai}	antenna temperature of station i
T_{bb}	temperature of black body $/K$

$T_{1,2}$	telescopes (stations, elements) $1, 2$
T_s	temperature of the source $/K$
T_{Si}	noise temperature of system (i) $/K$
$T(u)$	brightness (temperature) distribution over u
ΔT	time tolerance $/s$
T_k	tapes weight (assigned to the k-th visibility point)
t	time variable $/s$
t_a	antenna temperature spectrum
t_c	correlation time $/s$
t_s	Fourier transform of T_s
$t_s(s)$	spatial spectrum of the radio source
$U(f)$	Fourier spectrum of $u(t)$
$\underline{U}(u,t)$	envelope of the modulation
$U_{1,2}$	input voltage of an amplifier $/V$
U_{02}, V_{04}, V_{05}	output voltage of an amplifier $/V$
$U(t)$	open circuit voltage $/V$
u	universal argument (diffraction theory (4)), spatial frequency variable
$\underline{u}(t)$	complex modulation function
$u_0(t)$	output voltage of the system $/V$
u, v	$u - v$-plane coordinates (inverse, local positions)
u, v, w	coordinates in a right-handed coordinate system
$u_{\max} v_{\max}$	inverse of the EW and NS direction
V	visibility; brightness distribution of a source, amplification factor
V_a	interference factor
$V_i(t)$	voltage of station i
$V(d)$	visibility
$V(u)$	one-dimensional visibility function
$V(u,v)$	two-dimensional visibility function
$V(u,f)$	brightness distribution of the source
$\langle V_i \rangle$	time average of voltage of station i
$V_{ij,\text{obs}}(t)$	visibility, observed
$V_{ij,\text{true}}(t)$	visibility, true
$V'(u,v)$	weighted visibility function
$V''(u,v)$	reconvoluted visibility function
W	geometrical delay
W_{3db}	half angular width of the visibility function
W_s	totally radiated power $/W$
w	angular extension of a radio source
x	cartesian coordinate, variable quality
x_i	one-bit representation
y	cartesian coordinate
Z_0	free space impedance (377. Ohms)
z	cartesic coordinate

Greek symbols

α	phase of source brightness $/°$
α_0, δ_0	equatorial coordinates
α_R	right ascension
$\alpha_{a,b}, \delta_{a,b}$	coordinates of sources A, B
β	phase shift
$\Gamma(u_2, u_2, \tau)$	lateral coherence function (lcf)
γ	phase delay, source (mutual) coherence function, cross-spectral function
γ_L	loop gain
γ_N	normalized coherence function; degree of coherence
$\gamma(l_1, l_2, \tau)$	source coherence function
$\Delta\vartheta_\tau$	angular resolution in direction of the projected baseline
$\Delta\vartheta_{ff}$	fringe frequency angular resolution (HPBW) $/°$ or $','' $
$\Delta\vartheta_{f,\min}$	maximum angular resolution $/°$
$\Delta\vartheta_3$	half power beam width (HPBW) of the main lobe of an interferometer $/°$
$\Delta\kappa$	modulation index of the pseudo-noise code
$\Delta\vartheta$	retardation
$\Delta\delta$	declination error
$\Delta\tau$	correlation interval; sum of geometrical and systems delay
$\Delta\Phi_{a,b}$	phase shift appropriate to MASER A, B
$\Delta\Phi_{fa}$	phase shift
$\Delta\varphi$	modulation index of the pilot tone
$\Delta\Omega$	spatial angle (of a radio telescope) $/\text{sterad}$
δ	declination $/°$, baseline coordinate
δ_0	source declination
$\delta(t), \delta(u - u_k, v - v_k)$	Dirac's delta function
$\delta\varphi$	phase difference between incoming plane waves
ε	atmospheric refraction for small B
ε'	atmospheric refraction for large B
$\varepsilon(f)$	noise term
$\varepsilon(m, l)$	residual map
η	depolarisation error, efficiency $/\%$
Θ	angular resolution $/°$ or $','' $; angle perpendicular to $\vec{D}$
$\Theta_{i,j}$	phase of complex visibility function
Θ_m	phase
$\Theta_{m,n}$	phase changes between stations m, n
ϑ	zenith or polar angle $/°$
ϑ_0	axis of the $i.$ main lobe $/°$
ϑ'	elevation angle $(= 90° - \vartheta)$ $/°$
$\dot{\vartheta}$	time derivation of ϑ
λ	free space wavelength $/\text{cm}$
λ_0	carrier wavelength $/\text{cm}$
μ_0	free space permeability

ξ, ξ_0	angular variables, directions, transformed coordinate $/°$
$\xi = \sin \vartheta$	angular variables, directions, transformed coordinate $/°$
$\xi' = \sin \xi'$	angular variables, directions, transformed coordinate $/°$
$\rho_{1,2}(\tau)$	associated one-bit representation of the digitized signal
$\dot{\rho}$	angular velocity of the earth
ρ	complex crosscorrelation function (ccf)
$\vec{\rho}_{1,2}$	radius vectors to the i. elements no $1, 2$
σ	radar crosssection (rcs) of a radar target
σ_v	standard deviation of noise
$\sigma\varphi$	phase error
τ	general delay, time shift, transmission time to the target and back $/s$
τ_0	a priori geometrical delay
τ_0'	delay of signal of antenna
$\tau(t)$	group delay
τ_{atm}	atmospheric delay (of earth)
τ_{RB}	delay by earth rotation
$\dot{\tau}_0$	derivation of geometrical delay
$\dot{\tau}_g$	delay rate
τ_g	geometrical time delay $/s$
τ_i	system instrumental geometrical phase shift, time delay $/s$
τ_M	a priori mode delay
$\omega\big((u/\Delta u),(v/\Delta v)\big)$	Bracewells (293) cha function
Φ	phase, phase difference
Φ_B	phase due to instrumental properties and of fringe washing function
Φ_g	geometrical (natural) fringe rate
$\Phi_{m,n}$	phase changes between stations m, n
$\Phi_{v(1,2)}$	phase of visibility
$\Phi_{1,2}$	phase(s) at the correlator (input)
$\dot{\Phi}_{1,2}$	fringe rate
φ	azimuth angle, difference angle
$\varphi(t)$	pseudo-noise code
χ_2	Woodward's ambiguity function (af)
χ	predicted visibility
$\Psi(f)$	Fourier spectrum of the complex (time dep.) signal $\Psi(t)$
ψ_i	phase of the visibility function (chapt.: 7.9.3 11.2) $/°$
Ω_A	equivalent spatial angle (of an antenna (pattern))
ω	angular velocity; dH/dt
ω_B	angular velocity of earth
ω_e	intermediate angular frequency
ω_{IF}	pilot tone frequency

1. INTRODUCTION

Interferometry is a means of increasing *angular resolution* (18,25,44,80). We use resolution in the sense of the ability to distinguish two objects which are adjacent to one another. We can distinguish the objects in a number of ways:

1. spectrally
2. by angle
3. by distance
4. by velocity.

A precondition in all these cases is that sufficient signal power is received from the objects, or, as radio astronomers and radar engineers put it, the signal must stand out above the noise level. Consideration of the ambiguity function will however show that this condition need not always be strictly fulfilled. The prime purpose of interferometry is the achievement of high angular resolution, a topic which is dealt with at length in this book and the topic of *velocity* resolution is also discussed. The two other types of resolution mentioned above are briefly defined here. The *spectral* resolution (4) is the ratio of mean wavelength to the smallest wavelength difference $(\lambda_m / \Delta\lambda)$ of spectral lines which can just be resolved by a spectrometer: in the case of the sodium D lines at $\lambda_m = 600.nm$ this is 1000. The *distance* resolution Δr is dependent in the case of a radar system on the velocity of light and the resolvable time difference Δt (78):

$$\Delta r = (1/2)c.\Delta t \tag{1.1}$$

The factor $1/2$ is a consequence of the double path travelled by the transmitted and reflected signal. The *angular* resolution or resolution of an imaging device (4), for example a microscope, is defined by the minimum separation of two objects which are to be seen as separate images

$$d = \lambda_m / A ; \quad A = n.\sin\vartheta_0 . \tag{1.2}$$

Perpendicular incident rays are assumed and A is the *"numerical aperture"* which is the product of refraction index n between object and objective lens and the sine of the instrument acceptance angle.

The angular resolution of a radio telescope (2,4,24) is the ability to distinguish the diffraction images of two equal adjacent point sources when the angular distance between the sources is greater than the half-power width of the antenna main beam. The accuracy of measurement of an *angular frequency* (32,43) is also determined by the angular resolution. The *pointing accuracy* (24) of an antenna may however be determined with an accuracy up to two orders of magnitude better than this. Angular resolution and half-power beamwidth designate the same quantity and are determined by the antenna aperture expressed in free-space wavelengths. This corresponds to the *Rayleigh criterion* (4.12). For an aperture which is large in terms of wavelengths, a simple relation sufficiently accurate in practice can be given for angular resolution.

$$\vartheta(\text{rad}) \sim \lambda/D \text{ or } \vartheta(') \approx 3438/D , \tag{1.3}$$

where λ is the free space wavelength and D is the length of an extended aperture or its diameter (the letter D stands for *baseline* as will be seen later, equations (6.1), (6.6). In order to achieve an angular resolution of one minute of arc the size of this aperture or antenna array must be 3438λ. Table 1.1 is intended to give a feeling for the orders of magnitude of the angular resolution of various instruments

Angular Res. ($''$)	Kind of Instr.	Freq. f(GHz)	Wavel. (cm)	Diam./Baseline D(m)
60.	antenna	1.42	21.	722.
60.	antenna	0.10	300.	10314.
60.	naked eye	6.10^7	5.10^{-7}	$1.17\ 10^{-3}$
1.	opt. telesc.	6.10^7	5.10^{-7}	0.1

Table 1.1. Instrumental diameter for various angular resolutions.

To improve the angular resolution we know from optics that it is possible to use the *interferometer principle*: It says that for an optical telescope one does not need to use a filled aperture of size D in order to obtain the angular resolution corresponding to that aperture.

It may be sufficient to use only two small sections of the whole large aperture, separated by an appropriate distance. In this case the focal image of a point source of radiation is no longer transformed into a circularly symmetric *diffraction disc* or *Airy zone(s)* by diffraction effects but gives rise to a parallel system of *interference fringes*. The system of fringes can easily be used to determine the exact position of the radiation soure or to distinguish the two components of a double source. If an improvement of the angular resolution is required the spacing D between the apertures simply must be enlarged. It is therefore not necessary to increase the size of the optical telescope but merely to direct the received rays by means of mirrors into the original telescope optics where a focal plane image is generated in the usual way. The spacing D between the apertures determines the angular resolution of this so-called *interferometer*. This type of optical interferometer was employed in 1922 by Michelson as a *stellar interferometer* in order to extend the 2.5 m diameter reflector of the Mount Wilson observatory to an instrument with an effective aperture of 6. m (Fig. 1.1). With this ingenious device extremely close double stars could be resolved and even diameters of stars could be measured. With

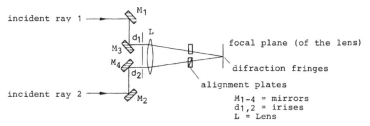

Fig. 1.1. Michelson's stellar interferometer $L =$ Lens.

the advent of radio astronomy it was natural to attempt to build such an instrument for the radio wavelength range. The two small apertures become identical antennas or antenna systems connected by cables of equal length to a receiver so that the signals received by the antennas are superposed with equal phase at the receiver. The use of this scheme of equal antennas leads to a *multi-lobe* antenna pattern and hence a high angular resolution. The receiver system is so narrowband that the incident radiation can be considered *monochromatic*. The broadband components received by the antenna are rejected by the receiver. Increasing the bandwidth increases the sensitivity of the system but leads to smearing of the multi-lobe pattern and a reduction of the resolution. Thus an interferometer using phase comparison has a resolution which is dependent on the necessarily narrow bandwidth used. An additional difficulty is that of performing the phase comparison when the baseline D is long, which prevents the use of arbitrarily large D. As in optics an interferometer at radio wavelength must preserve coherence of the two received signals, i.e. they must be capable of interfering with each other. The quantities *coherence time* and *coherence length* are relevant in this connection. The limited coherence of an incident wave packet is a consequence of the quantum nature of the radiation mechanism (quantum jump in an atom or molecule). The time interval is about $T = 10^{-8}s$ for visible light so using $T = L/c$ the coherence length L of the wave packet is only a few meters.

Using the basic equation for a twin interferometer (6.1) we can consider the following: for a given intensity in the interferometer or fringe pattern the variation $\delta\vartheta$ of the zenith angle and the path length variation $\delta\Delta l$ are connected by the relation(s)

$$\delta\Delta l = -2D(\cos\vartheta)\delta\vartheta \tag{1.4}$$

which fixes the allowable differences of path lengths from the receiving antennas to the common detector for a given uncertainty in the value of $\delta\vartheta$. If for example $\delta\vartheta$ is less than $10''$ or 5.10^{-5} radians for sources near the zenith ($\vartheta \approx 0$) then the path difference $\delta\Delta l$ must remain smaller than $10^{-4}D$. At $D = 1000.\lambda$ and $\lambda = 3.m$ then $\delta\Delta l \leq 0.3\ m$. If ϑ deviates further from zero, e.g. $20°$ then the restriction becomes greater according to the cosine law in equation (1.4). Various methods which have been developed to alleviate these disadvantages will be discussed in Chapter 7.

Following the first efforts of the Cambridge group in 1952 (97,148) to combine two antennas to make an interferometer these instruments found widespread use for radioastronomical measurements. In the intervening years both the instruments themselves and the associated analysis methods have been developed to a high degree of perfection. At first the radio sources observed were the sun and galactic objects, later the observations were extended to quasars and other much fainter objects.

The first applications of interferometry were in optical and radioastronomy, but subsequently the method was applied to a number of problems in which precise measurements of angle of incidence were required. The attainable resolution is however limited by effects and variations in the troposphere and ionosphere and in interplanetary and interstellar media, as presented in Fig. 5.3 (51). The high angular accuracy required in space applications led to the use of interferometry in radar ranging and in tracking satellites. The US minitrack network uses twenty interferometer ground stations to track various satellites and spacecrafts for instance the Gemini series. Interferometry can also be used to investigate the surface accuracy of large parabolic antennas.

2. CHARACTERISTIC MAGNITUDES OF COSMIC RADIO RADIATION

The main characteristics of cosmic *radio sources* can be expressed by the intensity and the polarization of the electromagnetic waves radiated. These magnitudes vary with frequency, the *angle of incidence* and time. The variation of the *intensity* with *frequency* is referred to the *spectrum*. Usually three magnitudes are used to express the intensity of cosmic radio noise: *brightness* or *brightness temperature* B'' for the measurement of the intensity of extended sources or of a diffuse *background* and *flux density* for the measurement of the intensity of the radiation of discrete sources.

It is assumed that the *energy* ΔE in a frequency interval f to $f + \Delta f$ during the time interval Δt perpendicularly on a small aperture of area ΔA. The radiation is incident within the solid angle $\Delta \Omega$. Thus the brightness may be expressed as

$$B'' = E/(\Delta \Omega \Delta A \Delta f \Delta t) . \tag{2.1}$$

The unit of brightness is Watt/(m^2.Hz.steradian). This brightness definition is independent of the distance between the observer and the radiating area (15,44 p. 13).

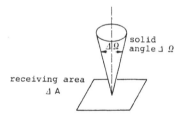

Fig. 2.1. The definition of the surface brightness.

As the intensity is often described in terms of a *noise temperature* it is often useful for comparison purposes to give the radiation of an extended source as a temperature. Then the radiation density of the source is set equal to that of a black body (4,27) of the same solid angle Ω and for the same frequency interval f to $f + \Delta f$. We set the temperature of the black body equal to the temperature T_s of the source radiation. Using the Rayleigh-Jeans limit of Planck's radiation law the brightness of a black body at a certain f is directly proportional to the temperature

$$B''(f) = 2kf^2T_{bb}/c^2 = 2k T_{bb}/\lambda^2 \tag{2.2}$$

where

T_{bb} = absolute temperature of the black body $(bb)(K)$,

k = Boltzmann's constant $(1.38.10^{-23} \text{Joule}/K)$,
c = velocity of light $(3.10^{10} \text{cm}/s)$,
λ = wavelength in free space.

The radiation of a black body is *randomly polarized* and the brightness $B''(f)$ refers to the total power contained in orthogonal polarization planes (87,88). Since then the total brightness of randomly polarized radiation is twice as large as that of completely polarized radiation, whether linearly, circularly or elliptically polarized, it follows from (2.2) for the case of completely polarized radiation that

$$B_c''(f) = k.f^2 T_{bb}/c^2 .$$ (2.3)

Thus we see that the brightness temperature of fully polarized radiation is proportional to brightness and independent of the polarization. For polarized radiation the total brightness can be regarded as a sum of two completely polarized, orthogonal components. To describe the radiation two temperatures are necessary, one temperature being associated with each polarization. If the radiation source has a well-defined extent, then it can be designated as a "discrete" source. The intensity S of such a source may be described by the integral of the brightness over the angular extent of the source.

$$S = \iint B'' \, d\Omega .$$ (2.4)

S is also referred to as the flux density (23,31,41). Its unit in the MKS system is Watts/(m^2.Hz) or Jy (Jansky, where one Jansky $= 10^{-26}$.W/m^2 Hz). S decreases with the inverse square of the distance between the observer and the radiating source.

3. RECEPTION OF QUASI-MONOCHROMATIC, PARTIALLY POLARIZED PLANE WAVES

An antenna is situated at the origin (15,87) of a spherical coordinate system (r, ϑ, φ). A partially polarized, *quasi-monochromatic*, plane wave impinges from the direction ϑ, φ on the antenna aperture where $\underline{\vec{E}}_e(r,t)$ corresponds to the complex field vector of the incident wave

$$\underline{\vec{E}}_e(r,t) = \underline{E}_\vartheta \, \vec{e}_\vartheta + \underline{E}_\varphi \, \vec{e}_\varphi \tag{3.1}$$

with

$$\underline{E}_\vartheta = A_1(t) \exp j\{\omega t + kr + a_1(t)\}$$

and

$$\underline{E}_\varphi = A_2(t) \exp j\{\omega t + kr + a_2(t)\} \ .$$

$\vec{e}_\vartheta$ and $\vec{e}_\varphi$ are unit vectors in the direction of the spherical coordinates, ω is equal to the average frequency and $k = \omega/c$. The open circuit voltage at the input terminals of the antenna can be derived either from the *reciprocity theorem* (8,18,26) or by using the complex *effective antenna height*

$$\underline{U}(t) = \underline{\vec{E}}_e \underline{\vec{h}} = \underline{E}_\vartheta \, \underline{\vec{h}}_\vartheta + \underline{E}_\varphi \, \underline{\vec{h}}_\varphi \ , \tag{3.2}$$

where

$$\underline{\vec{h}} = \underline{h}_\vartheta \, \vec{e}_\vartheta + \underline{h}_\varphi \, \vec{e}_\varphi \ .$$

If the same antenna is used for transmission, the far field $\underline{E}_T$ produced by an input current $\underline{I}_e$ is

$$\underline{E}_T(r, \vartheta, \varphi, t) = \underline{E}'_\vartheta \, \vec{e}_\vartheta + \underline{E}'_\varphi \, \vec{e}_\varphi =$$
$$= \left\{ -j\omega\mu_0 \underline{I}_e \underline{\vec{h}}(\vartheta, \varphi) \exp\left(j(\omega t - kr)\right) \right\}/4\pi r \tag{3.3}$$

where

$$\underline{E}'_\vartheta = B_1 \exp\{j(\omega t - kr + \beta_1)\}$$

and

$$\underline{E}'_\varphi = B_2 \exp\{j(\omega t - kr + \beta_2)\} \ ,$$

B_1, B_2, β_1 and β_2 are time-independent quantities on the assumption that the antenna radiates a *monochromatic* wave. Generally $\underline{E}_T$ is elliptically polarized. Further we assume that $\underline{h}$ is frequency independent within the small bandwidth Δf. For the directivity $D(\vartheta, \varphi)$ of the antenna the *radiation resistance* R_s can be expressed $\underline{\vec{h}}(\vartheta, \varphi)$ to obtain

$$D(\vartheta,\varphi) = \vec{E}_T\,\vec{E}_{T^*}/\left\{(1/4\pi)\iint \vec{E}_T\,\vec{E}_{T^*}\,d\Omega\right\} =$$

$$= \vec{\underline{h}}\,\vec{\underline{h}}^{*}/\left\{(1/4\pi)\iint \vec{\underline{h}}\,\vec{\underline{h}}^{*}\,d\Omega\right\} \tag{3.4}$$

and

$$R_s = 2W_s/|\underline{L}_e|^2 = (Z_0/4\lambda^2)\iint_{4\pi}\underline{h}\,\underline{h}^{*}\,d\Omega \;, \tag{3.5}$$

where W_s is the total radiated power, Z_0 the free space impedance and $d\Omega = \sin\vartheta\,d\vartheta\,d\varphi$, the solid angle element. The energy intercepted by the receiving antenna from the incident plane wave can be computed as (24,25)

$$W = \overline{U\,U^{*}}/8R_s \;, \tag{3.6a}$$

where the bar indicates the *time average* of the general form

$$\overline{f(t)} = \lim_{T\to\infty}(1/2T)\int_{-T}^{+T} f(t)\,dt \;. \tag{3.6b}$$

Introducing (3.2) in (3.6) yields

$$W = \overline{(\vec{E}_e\,\vec{\underline{h}})(\vec{E}_e\,\vec{\underline{h}})}{}^{*}/(8R_s) = (1/8R_s)\overline{(E_\vartheta\,\underline{h}_\vartheta + E_\varphi\,\underline{h}_\varphi)}.$$

$$\overline{(E_\vartheta^{*}\,\underline{h}_\vartheta^{*} + E_\varphi^{*}\,\underline{h}_\varphi^{*})} = (1/8R_s)\Big(\underline{h}_\vartheta\,\underline{h}_\vartheta^{*}\,\overline{E_\vartheta\,E_\vartheta^{*}} + \underline{h}_\vartheta\,\underline{h}_\varphi^{*}\,\overline{E_\vartheta\,E_\varphi^{*}} +$$

$$+ \underline{h}_\vartheta^{*}\,\underline{h}_\varphi\,\overline{E_\vartheta^{*}\,E_\varphi} + \underline{h}_\varphi\,\underline{h}_\varphi^{*}\,\overline{E_\varphi\,E_\varphi^{*}}\Big) \;. \tag{3.7}$$

Introducing equations (3.4) and (3.5) in (3.7) yields

$$W = \frac{\lambda^2 D}{4\pi}\,.P.\,\frac{\Big(\underline{h}_\vartheta\,\underline{h}_\vartheta^{*}\,\overline{E_\vartheta\,E_\vartheta^{*}} + \underline{h}_\vartheta\,\underline{h}_\varphi^{*}\,\overline{E_\vartheta\,E_\varphi^{*}} + \underline{h}_\vartheta^{*}\,\underline{h}_\varphi\,\overline{E_\vartheta^{*}\,E_\varphi} + \underline{h}_\varphi\,\underline{h}_\varphi^{*}\,\overline{E_\varphi\,E_\varphi^{*}}\Big)}{\Big(\overline{E_\vartheta.E_\vartheta^{*}} + \overline{E_\varphi.E_\varphi^{*}}\Big)(\underline{h}_\vartheta.\underline{h}_\vartheta^{*} + \underline{h}_\varphi.\underline{h}_\varphi^{*})} \;, \tag{3.8}$$

where

$$P = \Big(\overline{\vec{E}_\varphi\,\vec{E}_\varphi^{*}}\Big)/(2Z_0)$$

which corresponds to the total *power density* of the incident wave. As this indicent wave is assumed only partially polarized, its field components $\underline{E}_\vartheta$ and $\underline{E}_\varphi$ can be represented by the sum of a term with randomly distributed polarization and a completely polarized, coherent term:

$$\underline{E}_\vartheta(\vartheta,\varphi) = {}_s\underline{E}_\vartheta(\vartheta,\varphi) + {}_c\underline{E}_\vartheta(\vartheta,\varphi) \;, \tag{3.9a}$$

$$\underline{E}_\varphi(\vartheta,\varphi) = {}_s\underline{E}_\vartheta(\vartheta,\varphi) + {}_c\underline{E}_\varphi(\vartheta,\varphi) \;. \tag{3.9b}$$

The indices s and c refer to stochastic and coherent terms. Further, the following relations are assumed valid

$$\overline{{}_s\underline{E}_\vartheta\,{}_c\underline{E}_\vartheta^{*}} = \overline{{}_s\underline{E}_\varphi\,{}_c\underline{E}_\varphi^{*}} = \overline{{}_s\underline{E}_\vartheta\,{}_s\underline{E}_\varphi^{*}} = 0$$

$$\overline{{}_s\underline{E}_\vartheta\,{}_s\underline{E}_\vartheta^{*}} = \overline{{}_s\underline{E}_\varphi\,{}_s\underline{E}_\varphi^{*}} \;.$$

Hence the fractional polarization p (15,23) or *polarization ratio* (26, p. 76) can be obtained

$$p = \left(\overline{_cE_\vartheta \, _cE_\vartheta^*} + \overline{_cE_\varphi \, _cE_\varphi^*}\right) \Big/ \left(\overline{E_\vartheta \, E_\vartheta^*} + \overline{E_\varphi E_\varphi^*}\right) . \tag{3.10}$$

Substituting equation (3.9) in (3.8) and using equation (3.10) yields the intercepted energy.

$$W = (1/2)(1-p)A_e \, P + p\eta \, P.A_e \tag{3.11}$$

where

$$\eta = \frac{h_\vartheta \, h_\vartheta^* \, \overline{_cE_\vartheta \cdot _cE_\vartheta^*} + h_\vartheta h_\varphi^* \, \overline{_cE_\vartheta \cdot _cE_\varphi^*} + h_\vartheta^* \, h_\varphi \cdot _c\overline{E_\vartheta^* \cdot _cE_\varphi} + h_\varphi^* \, h_\varphi \, \overline{_cE_\varphi \cdot _cE_\varphi^*}}{\left(\overline{_cE_\vartheta \, _cE_\vartheta^*} + \overline{_cE_\varphi \cdot _cE_\varphi^*}\right)(h_\vartheta \, h_\vartheta^* + h_\varphi \, h_\varphi^*)} \tag{3.12}$$

and

$$A_e = \lambda^2 \, D/(4\pi) . \tag{3.13}$$

A_e corresponds to the effective area or aperture of the antenna. The *polarization efficiency* η is only determined by the completely polarized part of the incident wave and the polarization characteristics of the antenna, and it can be shown that η lies in the range $0 \le \eta \le 1$.

The first term in (3.11) represents the relation between the aperture of the antenna system and the randomly polarized part of the incident wave; it is independent of the polarization properties of the antenna itself.

The second term, on the contrary, stands for the completely polarized part and depends on the antenna polarization. It varies with η from 0 to $p \, P \, A$ where η is called the *depolarization factor* (23,40) since it describes the loss due to polarization mismatch between the antenna and the incident wave.

4. RECEPTION OF RANDOMLY POLARIZED WAVES FROM AN EXTENDED, INCOHERENT SOURCE

Here we derive the relation between the antenna and the microwave radiation coming from an incoherent (15,87) extended source (e.g. a gas cloud). For the sake of simplicity the incident wave is assumed to be randomly polarized and confined to the frequency range f to $f + \Delta f$. The radiation is described by its brightness $B''(\vartheta, \varphi)$. The energy intercepted by unit area is given by $B''(\vartheta, \varphi)d\Omega\, df$. The energy per unit bandwidth (15,87) received by an antenna of effective aperture area A_e from the incident wave is

$$dW = (1/2)\, A_e(\vartheta, \varphi)\, B''(\vartheta, \varphi)d\Omega \ . \tag{4.1}$$

Since in our case the source is incoherent, the total energy received by the antenna is equal to the sum of energies of the components of the source over its solid angle (27,31) and

$$W = \iint_\sigma dW = (1/2) \iint_\sigma A_e(\vartheta, \varphi)\, B''(\vartheta, \varphi)d\Omega \ . \tag{4.2}$$

Often it is convenient to express the noise energy received by the antenna in terms of the *equivalent thermal noise* of a matched real resistance.

$$W = k\, T_a \ . \tag{4.3}$$

T_a is antenna noise temperature or simply the antenna temperature. We derived brightness in a similar way using temperature equivalence in equation (2.2).

Substituting (2.2) and (4.3) into equation (4.2) we obtain

$$T_a(1/\lambda^2) \iint A_e(\vartheta, \varphi)\, T_s(\vartheta, \varphi)d\Omega \ , \tag{4.4}$$

where $T_s(\vartheta, \varphi)$ is the *brightness temperature of the source*. The aperture area A_e can also be written as

$$A_e(\vartheta, \varphi) = A_m \cdot f(\vartheta, \varphi) \tag{4.5}$$

where A_m is the maximum value of A_e and $f(\vartheta, \varphi)$ the normalized *power pattern* of the antenna (8,21,23,40) and $f(\vartheta, \varphi)$ is defined in the region $0 \le f(\vartheta, \varphi) \le 1$. The integration of $f(\vartheta, \varphi)$ over the surface of the *unit sphere* yields the *equivalent solid angle* Ω_a of the antenna characteristic

$$\Omega_a = \iint_{4\pi} f(\vartheta, \varphi)d\Omega \ . \tag{4.6}$$

We further assume that

$$\Omega_a \cdot A_m = \lambda^2 \tag{4.7}$$

is valid (18,23), so that the antenna temperature T_a can be rewritten as

$$T_a = (1/\Omega_a) \iint f(\vartheta, \varphi)\, T_s(\vartheta, \varphi)d\Omega \ . \tag{4.8}$$

The derivation of the above equations was made on the basis of a randomly polarized incident wave but it is evident that the results can also be valid for the case of partly polarized waves (27,40).

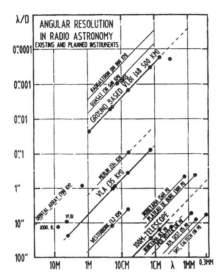

Angular resolution in radio astronomy (313).

5. THEORY OF LINEAR SPATIAL FILTERS

The usual method of determining the *brightness* or intensity distribution of a section of the sky is to make a raster scan of it using a single antenna with a *pencil beam*, in the same way as a television picture is scanned. As the output of such an antenna can be displayed on a TV screen or otherwise, we can consider the antenna as an *image dissector* which translates the angular distribution of radio brightness into a picture or diagram which allows the astronometer to deduce as much as possible about the true *brightness distribution* of the source. It is important to know the accuracy with which the antenna can produce a radio picture of the object in question. Bracewell and Roberts (36,89) were the first to investigate this question using the idea of *spatial frequency*, performing spectral analysis by means of the Fourier transformation (71,72,73,89). Interferometric observations are based on the van Cittert-Zernike theorem (5), which defines the complex degree of *coherence* $\mu_{1,2}$ between a fixed point P_2 and a movable point P_1 in a plane illuminated by *quasi-monochromatic* vibrations from the source. This quantity is equal to the normalized magnitude at P_1 in a *diffraction* centered at P_2 obtained by replacing the source with a diffraction aperture of the same size and shape and filling it with a spherical wave converging to P_2, the amplitude distribution over the aperture being proportional to the *intensity distribution* across the source.

$$\mu_{1,2} = e^{i\psi} \int_S\!\!\int I(x,y)e^{-ik(px-qu)}dxdy \, / \int_S\!\!\int I(x,y)dxdy \tag{5.1a}$$

where ψ is the phase factor of a two-dimensional circular source and p and q are dimensionless geometry factors describing the *projection* of the interferometer *baseline* on the $x - y$ plane (the sky plane). Introducing dependence on time (t) we can define the *lateral coherence function* (l.c.f.) or *visibility* in the time interval $\pm T$

$$\gamma(u_1, u_2, \tau) = \lim_{T\to\infty} (1/2T) \int_{-T}^{+T} V(u_1,t).V^*(u_1, t - \tau)dt \tag{5.1b}$$

which corresponds to a *convolution* of the voltages V, V^* with the geometric variables u.

The van *Cittert-Zernike Theorem* for the relation between the *coherence function* $c(\vec{b}, t)$ and the *brightness distribution* $I(\vec{\sigma})$ of an *extended* source may be expressed in a less complicated form if two boundary conditions are fulfilled (46, p. 164, 47, p. 260, Verschuur)

1. Source in the far field of the antenna
2. Integration time of the receiver less than the inverse of the bandwidth.

The relation is:

$$c(\vec{b}, \tau) = \int\!\!\int_\sigma I(\vec{\sigma}).\exp(j\, 2\pi\, \vec{b}^*.\vec{b})d\vec{\sigma} \ . \tag{5.1c}$$

The vectoral, three-dimensional relation may be separated into separate scalar relations for each of the three individual angular coordinates, e.g. spherical coordinates: r, ϑ, φ; here only the ϑ-dependence is considered.

For the sake of simplicity we may imagine that the power response and the *brightness distribution* of an incoherent randomly polarized object is only a function of elevation. The antenna points in the direction ϑ_0, and a point source of unit intensity is situated in the direction ϑ_s. The output power of the antenna can be written by $f(\vartheta_0 - \vartheta_s)$, and can be written as:

$$f(\vartheta_0) = \int\limits_{-\pi}^{+\pi} f(\vartheta_0 - \vartheta_s)\delta(\vartheta_s)d\vartheta_s \ , \qquad (5.1d)$$

where δ is the *Dirac delta* function.

$f(\vartheta_0)$ is normalised so that

$$\int\limits_{-\pi}^{+\pi} f(\vartheta_0)d\vartheta_0 = 1 \ . \qquad (5.2)$$

Let $T_s(\vartheta_0)$ be the sky brightness distribution. Radiation coming from different directions is assumed incoherent; if the antenna main beam points in the direction ϑ_0, the observed antenna temperature $T_a(\vartheta_s)$ will be

$$T_a(\vartheta_0) = \int\limits_{-\pi}^{+\pi} f(\vartheta_0 - \vartheta_s)T_s(\vartheta_s)d\vartheta_s \ . \qquad (5.3)$$

This antenna temperature is an exact mean value of the sky radiation temperature within the *antenna beam*.

If the antenna has a narrow pencil-shaped main beam, $f(\vartheta_0)$ is only determined by ϑ_0 within a small range, and is negliglible outside this range. In this case the integration limits for equation (5.3) can be increased to $\pm\infty$ (8). The observed brightness distribution can be adequately described by a *convolution* of the antenna beam with the brightness distribution, and we can write:

$$T_a(\vartheta_0) = f(\vartheta_0)^* T_s(\vartheta_0) \qquad (5.4)$$

using a *Fourier transformation* in the image plane

$$t_a(s) = F(s).t_s(s) \qquad (5.5)$$

where

$$t_s(s) = \int\limits_{-\infty}^{+\infty} T_s(\vartheta_0)\exp(-j2\pi s\vartheta_0)d\vartheta_0 \ . \qquad (5.6)$$

The quantity s is referred to as *"spatial" frequency*. If the real brightness distribution $T_s(\vartheta_0)$ is required then following (5.5) we can write the *spatial spectrum*

$$t_s(s) = t_a(s)/F(s) \qquad (5.7)$$

and using the *inverse Fourier transform*

$$T_s(\vartheta) = (1/2\pi) \int\limits_{-\infty}^{+\infty} t_s(s).\exp(+j2\pi s\vartheta_0)ds \ . \tag{5.8}$$

It is obvious from (5.5) that the *Fourier spectrum* of the observed brightness distribution can only contain Fourier components common to both the antenna beam and the true brightness distribution. Using the concept of spatial frequency one can consider the antenna as a *bandpass* which cuts off all spatial frequencies outside the desired range.

Similarly, the electric field distribution in the *far field* of a one-dimensional aperture may be expressed as a Fourier transform

$$\underline{E}(\sin \vartheta_0) = \int\limits_{-\infty}^{+\infty} f(x/\lambda)\exp\Big\{-\big((j2\pi x)/\lambda\big)\sin \vartheta_0\Big\}.d(x/\lambda) \tag{5.9}$$

and therefore

$$f(x/\lambda) = \int\limits_{-\infty}^{+\infty} \underline{E}(\sin \vartheta_0)\exp\{(j2\pi.x/\lambda)\sin \vartheta_0\}\, d(\sin \vartheta_0) \tag{5.10}$$

where the variable x/λ is the distribution of the electric field over the one-dimensional aperture. Using the *convolution theorem* (73) an expression of the following form may be obtained

$$f(x\lambda)^* f(x/\lambda) \simeq \int\limits_{-\infty}^{+\infty} f\{(x/\lambda)-\xi\}^* f(\xi)d\xi = \text{const.} \int\limits_{-\infty}^{+\infty} |\underline{E}(\sin \vartheta)|^2.$$

$$.\exp\{(j2x\pi \sin \vartheta_0)/\lambda\}\, d(\sin \vartheta_0) \ . \tag{5.11}$$

$|\underline{E}(\sin \vartheta_0)|^2$ is proportional to the *normalized power pattern* $f(\vartheta)$ of the antenna. If it is assumed that $f(\vartheta)$ or $\underline{E}(\sin \vartheta_0)$ can be estimated only for small values of ϑ, equation (5.11) may be written as follows

$$\int\limits_{-\infty}^{+\infty} |\underline{E}(\sin \vartheta_0)|^2 \exp\{(2j\pi x \sin \vartheta_0)/\lambda\}.d(\sin \vartheta_0) \propto f(\vartheta_0).$$

$$\exp(2j\pi x \vartheta_0/\lambda)d\vartheta \ . \tag{5.12}$$

Introducing $s = x/\lambda$ as a new variable in (5.11)

$$F(s) = \int\limits_{-\infty}^{+\infty} f(\vartheta_0)\exp(j2\pi s\vartheta_0)d\vartheta_0 \approx \text{const.} \int\limits_{-\infty}^{+\infty} f(s+\xi).f(\xi).d\xi \ . \tag{5.13}$$

The Fourier transform of $f(\vartheta_0)$ or the *spatial frequency filter characteristic* of the antenna is therefore approximately determined by the *autocorrelation function* (acf) of the aperture illumination. For an aperture of finite baseline length B this acf disappears at *spatial frequencies* of magnitude $s = x/\lambda$ which are larger than $s_L = D/\lambda$. In other words the antenna is not sensitive to spatial frequencies greater than the *limiting spatial frequency* s_L.

This statement is however not rigorously true since both equations (5.4) and (5.13) represent approximations, nevertheless for all practical purposes one can ignore spatial frequencies above s_L. Fig. 5.1 again compares the relation between time and frequency domains used in network theory with the analogous properties of antenna far field patterns and spatial frequency. Here h is the Dirac step response and C' the antenna power shifted by convolution.

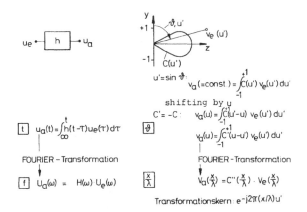

Fig. 5.1.   *Time (t)-angular (u') and frequency (f)-spatial*
frequency (x/λ) analogies.

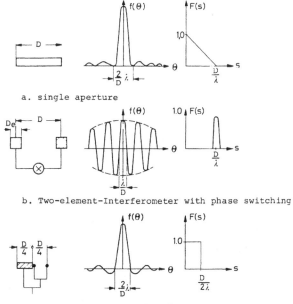

a. single aperture

b. Two-element-Interferometer with phase switching

c. Compound-Intensity Interferometer

Fig. 5.2.   *Interferometer arrangements, aperture form*
and spatial frequency spectrum
(continued next page).

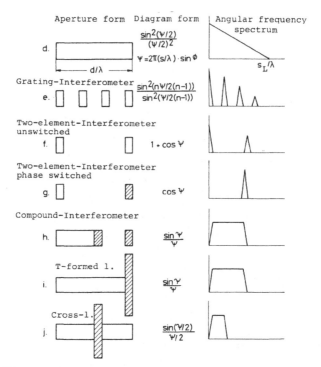

Fig. 5.2. Interferometer arrangements, form of aperture, form of power patterns, spatial frequency spectrum ("switched" interferometer elements are hatched).

Fig. 5.2 shows the aperture illumination, power pattern and spatial spectra obtained for seven practical antenna systems. The spectrum of Fig. 5.2a shows that a single aperture antenna has low-pass character. Spatial frequency components coming from the radiation source which are higher than the limiting value corresponding to aperture size are suppressed. The amplitudes of lower-frequency components are changed. This single aperture is compared to two interferometer configurations. Fig. 5.2g shows the corresponding diagrams for a phase-switched compound interferometer, which covers only a narrow band of spatial frequencies, whereas the compound interferometer of Fig. 5.2h has the spatial frequency response of an ideal low pass filter since it does not attenuate low frequencies. Chapter 7 deals with the various systems in more detail.

We shall now consider antennas arranged as a one-dimensional array (90). For such a configuration the complex signal voltage from the source of a fixed time is

$$\underline{U}(\beta \sin \vartheta_0) \exp(j\omega_0 t) = \underline{U}(u, t) \exp(j\omega_0 t) , \qquad (5.14)$$

where

t = time

ω_0 = angular frequency

$\underline{U}(u,t)$ = slowly varying modulation envelope of the signal from direction

u = $\beta \sin \vartheta_0$ with *Gaussian distribution* (over time)

β = phase constant.

Thus the *brightness temperature distribution* can be written as

$$T_s(u) = \overline{|\underline{U}(u,t)|}^2 .$$ (5.15)

T_s is real and proportional to the t average of the power from the direction u. If on the other hand the complex field $u(x_i,t)\exp(j\omega t)$ with $i = 1,2,...$ is considered at the positions x_1 and x, then the following complex *function* $f(x_1,x_2)$ may be defined

$$f(x_1,x_2) = \overline{\underline{u}(x_1 t)\underline{u}^*(x_2,t)} .$$ (5.16)

This function corresponds to the complex *cross-correlation function* (c.c.f.) with the time shift $\tau = 0$, (55,67,84). For incoherent sources this c.c.f. depends only on the mutual distance of both points $x = x_1 - x_2$, in this case the spacing of antennas. So the brightness $T_s(u)$ and $f(x_1,x_2) = f(x)$ may be represented

$$f(x) = \int_{-\infty}^{+\infty} T_s(u)\exp(jux)dx$$ (5.17)

and

$$T_s(u) = (1/2\pi) \int_{-\infty}^{+\infty} f(x)\exp(-jux) .$$ (5.18)

In this case $f(x)$ corresponds to the spatial frequency spectrum of $T_s(u)$ and equation (5.17) follows that by measuring $f(x)$ the true brightness distribution can be derived. Since the spatial frequency spectrum is band-limited ($T_s(u) = 0$ for $|u| > \beta$), the function $f(x)$ may be sampled in intervals of half a wavelength and taking into account the *sampling theorem* (67,73,75) we obtain

$$f(x) = \sum_{k=-\infty}^{+\infty} f(k\lambda/2).si\left\{\beta\left(x - (k\lambda/2)\right)\right\}$$ (5.19)

where the *si*-function is defined as $si(x) = (\sin x)/x$ and λ corresponds to the wavelength of the incident signal. Substituting equation (5.19) in (5.18)

$$T_s(u) = \begin{cases} \displaystyle\sum_{k=-\infty}^{+\infty} f(k.\lambda/2)\exp\{(jk\lambda u)/\lambda\} & \text{for } |u| \leq \beta \\ 0 & \text{for } |u| \geq \beta \end{cases}$$ (5.20)

or by variation of the summation limits

$$T_s(u) = \begin{cases} \displaystyle f(0) + 2\,\mathrm{Re}\left\{\sum_{k=1}^{+\infty} f(k\lambda/2)\exp\{(jk\lambda u)/2\}\right\} & \text{for } |u| \leq \beta \\ 0 & \text{for } |u| \geq \beta \end{cases}$$ (5.21)

as $f(x) = f(-x)$, if $T_s(u)$ is real.

If the spatial frequency spectrum is measured in intervals of $\lambda/2$ for $x \geq 0$, then the brightness temperature distribution may be reconstructed. The accuracy of the result is governed by the number of samples taken.

The spatial frequency spectrum therefore represents a useful way to describe the angular resolution of an antenna array system and is an important parameter for the description of the *imaging* characteristics of a single antenna element or interferometer array. Limits of this resolution are shown in Fig. 5.3.

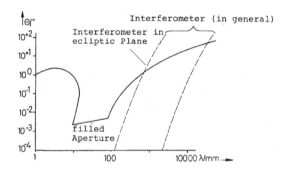

Fig. 5.3.   *Limit of radio interferometer angular resolution after (51)*

6. INTERFEROMETRIC TECHNIQUES

6.1. Principle of the Interferometer

We consider two identical, fixed, *isotropically* radiating antennas separated by an east-west baseline of $D = n.\lambda$, both being connected to a receiver (Fig. 6.1), where λ is the wavelength (9,22,26,60).

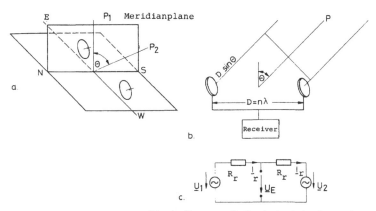

Block diagram of the twin interferometer

Fig. 6.1. The principle of the twin interferometer

The beam of the system is formed by *multiplication* (15,18,21,32) of the beam of the single element $\underline{L}(\vartheta)$ with the *array pattern* $\underline{M}(\vartheta)$. The array pattern is responsible for multilobe structure while the element pattern dictates the shape of the envelope as a whole, as shown in Fig. 6.2a. The power delivered to the receiver is proportional to $|\underline{L}(\vartheta).\underline{M}(\vartheta)|^2$. Since the lobe structure is of interest for an interferometer we can assume that the array pattern is determined solely by $\underline{M}(\vartheta)$ and set $\underline{L}(\vartheta)$ to unity. Although this is not true for real radioastronomical applications it is valid for the case in which radiation is received by the central lobes of the array pattern (22). If we now use the array to observe a point source in the *meridian plane*, i.e. perpendicular to the *baseline vector* $\vec{D}$ then the parallel rays from the source arrive in phase at both antenna elements. If the signals are transferred by cables of equal length to the receiver both will arrive there in phase and will be superposed.

If the source is fixed in space, earth rotation will move it gradually out of the field of view. The rays are then no longer incident in phase at the eastern and western elements. The sine of the angle of *incidence* of the rays is equal to the path length difference between east and west antennas by the baseline length $D = n\lambda$. Thus the *phase difference* Φ is (Fig. 6.1b)

$$\Phi = (2\pi D/\lambda) \sin \vartheta \; . \tag{6.1}$$

The signals which now propagate along the cables are phase shifted by Φ and the total power to the receiver will be lower (Fig. 6.1b). The voltages induced in the antennas are

$$\underline{U}_1(\vartheta) = \vec{\underline{E}}_e \, \underline{\vec{h}}(\vartheta) \exp(+j\Phi/2)$$

and

$$\underline{U}_2(\vartheta) = \vec{\underline{E}}_e \, \vec{h}(\vartheta) \exp(-j\Phi/2) \; . \tag{6.2}$$

$\vec{\underline{E}}_e$ corresponds to the electric field of the source at the antenna element, and $\underline{\vec{h}}(\vartheta)$ is the *effective vectorial* complex *height* of the appropriate element. The open circuit voltage at the receiver can be found using the *equivalent circuit* of Fig. 6.1c.

$$\underline{U}_E = \underline{U}_1 - \underline{i}R_r = \underline{U}_2 + \underline{i}R_r$$

$$\underline{U}_E = (\underline{U}_1 + \underline{U}_2)/2 \; , \tag{6.3}$$

where R_r is the radiation resistance of one antenna of the interferometer. The power available in this system is

$$
\begin{aligned}
P(\vartheta) = (\underline{U}_E \, \underline{U}_E)/4R_r &= \{(\underline{U}_1 + \underline{U}_2)(\underline{U}_1 + \underline{U}_2)\}/(16R_r) = \\
&= \left(\{\vec{\underline{E}}_e \, \vec{\underline{h}}^2(\vartheta)\}/8R_r \right) . (1/2) \{\exp(+j\Phi/2) + \exp(-j\Phi/2)\} \{\exp(-j\Phi/2) + \\
&\quad + \exp(+j\Phi/2)\} = \\
&= P_0(\vartheta)(1/2) \{2 + \exp(+j\varphi) + \exp(-j\Phi)\} = \\
&= P_0(\vartheta)(1 + \cos \Phi) \tag{6.4}
\end{aligned}
$$

with

$$P_0(\vartheta) = \{\vec{\underline{E}}_e \, \vec{\underline{h}}^2(\vartheta)\}/(8R_r) \; . \tag{6.5}$$

Here $P_0(\vartheta)$ corresponds to the power available from a single element. For small ϑ the approximation $\sin \vartheta \approx \vartheta$ can be used and (6.4) is written as

$$P(\vartheta) = P_0(\vartheta) \{1 + \cos(2\pi D/\lambda)\} \; . \tag{6.6}$$

As mentioned above, this power P is zero if both signals are in antiphase, then the path difference is equal to half a wavelength and the angle by which the source has moved can be computed as

$$\Phi_{\lambda/2} = (2\pi D/\lambda) \sin \vartheta = (2\pi n\lambda/\lambda) \sin \vartheta = 2\pi n \sin \vartheta \; . \tag{6.7a}$$

Zeros of this function occur at angles of

$$\Phi = \pi + 2k\pi \quad (k \text{ is integer})$$

or

$$\pi + 2k\pi = 2\pi n \sin \vartheta \tag{6.7b}$$

and a series of angles $\vartheta_1, \vartheta_2 ... \vartheta_n$ are found which are solutions of this equation. Similarly, maxima of the received energy occur at

$$\Phi = 2m\pi = (2\pi n\lambda/\lambda)\sin\vartheta = 2\pi n\sin\vartheta \quad (n\ \text{integer})\ . \tag{6.8a}$$

Again considering small angles we write $\sin\vartheta \approx \vartheta$ and minima are found at

$$\pi + 2k\pi = 2\pi n\,\vartheta_{\min}$$

$$\vartheta_{\min} = \{k + (1/2)\}/n\ \text{in rad.}\ , \tag{6.8b}$$

and

$$2m\pi = 2\pi n\,\vartheta_{\max}$$

$$\vartheta_{\max} = m/n \tag{6.8c}$$

where

$$n = D/\lambda = Df/c\ . \tag{6.8d}$$

This integer number m, which fixes the position and number of maxima in relation to the central axis is called the *order of interference or fringe* (2,9), Fig. 6.2.

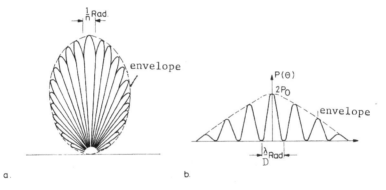

Fig. 6.2.   *Field strength and power pattern of a twin*
interferometer
a. Radiating pattern of a twin interferometer (polar)
b. Power pattern in cartesian coordinates.

These maxima and minima are regularly distributed around the meridian ($\vartheta = 0$.) and separated by an angle $(1/2)\,n^{-1}$ (Fig. 6.2a). In the course of the daily rotation of the earth the radiation source traverses the regular series of lobes of the interferometer with constant angular velocity and a recording of receiver ouput shows interference phenomena or fringes as shown in Fig. 6.2b. $P(\vartheta)$ varies in the interval $0 < P(\vartheta) < 2P_0(\vartheta)$. Strictly this is only valid for rather small values of ϑ. Outside the range of small ϑ the assumption that the absolute amplitude of the element pattern $|L(\vartheta)|$ is equal to unity must be dropped and the equation (6.4) has to be multiplied by $|L(\vartheta)|^2$, resulting in a modulation of the *interference amplitudes* (see Fig. 6.2b).

However, if the object is tracked with the central lobe of the beam the power pattern of the interferometer appears as shown in Fig. 6.3. As the angular dimension between two

neighbored minima is n^{-1} the *half power beam width* (HPBW) or the angular resolution
is $(1/2\,n)$ (radians). For this reason the angular resolution of an interferometer with the
baseline D is equal to that of an array of continuously distributed elements of total length
$2D$. The twin interferometer shown in Fig. 6.1a was first introduced in this simple form
by Ryle and Vonberg (96) in 1946 in England (9,19); see also section 7.9.3. This type
of interferometer, which is analogous to Michelson's, is a *meridian transit* interferometer.
Even if the angular resolution of a single lobe of the pattern is very high, in the presence
of multiple radiation sources difficulties occur as the pattern of interferometer lobes covers a relatively wide angular region. In this case the sources can no longer be *uniquely*
distinguished.

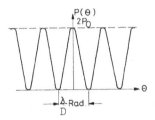

*Fig. 6.3.   Power characteristics of a twin interferometer
while tracking a source.*

This leads to the tendency to increase the angular resolution of each antenna of the interferometer by replacing the hypothetical isotropic (low gain) element by large reflector
antennas with high gain and small beamwidth.

6.2. Effect of Bandwidth

In the preceeding section it was assumed that the radio source is a point radiating mono-
chromatically, since according to the rules of wave optics only monochromatic radiation
is coherent and therefore capable of interference. Strictly speaking, this condition is not
fulfilled by any of the real physical sources, either in the optical range or at radio frequency
(RF). Even very narrow spectral lines (240) have a *finite bandwidth*. Therefore the signal
response of a simple twin interferometer to the radiation of a point source with bandwidth
$f_0 \pm (\Delta f/2)$ can be described by the following function of the angle variable ϑ

$$P(\vartheta) = \int_{f_0-(\Delta f/2)}^{f_0+(\Delta f/2)} P_0(1+\cos\Phi)df \tag{6.9a}$$

$$\Phi = (\omega D \sin\vartheta)/\lambda = 2\pi f D \sin\vartheta/c . \tag{6.9b}$$

Then

$$P(\vartheta) = \int_{f_0-(\Delta f/2)}^{f_0+(\Delta f/2)} P_0\{1+\cos(2\pi f D \sin\vartheta/c)df =$$

after substitution

$$P(\vartheta) = P_0 \left(f + \frac{\sin\{(2\pi fD \sin \vartheta)/c\}}{(2\pi D \sin \vartheta)c} \right) \Big|_{f_0-(\Delta f/2)}^{f_0+(\Delta f/2)} =$$

$$= P_0 \left\{ f_0 + \frac{\Delta f}{2} + \frac{\sin\left(((2\pi D f_0 \sin \vartheta)/c) + 2\pi \frac{D}{c} \frac{\Delta f}{2} \sin \vartheta\right)}{(2\pi D \sin \vartheta)/c} - \right.$$

$$\left. - f_0 + \frac{\Delta f}{2} - \frac{\sin\left(((2\pi D f_0 \sin \vartheta)/c) - 2\pi \frac{D}{c} \frac{\Delta f}{2} \sin \vartheta\right)}{(2\pi D \sin \vartheta)/c} \right\}, \qquad (6.9d)$$

and using the substitution

$$\sin a - \sin b = 2 \cos\big((a + b)/2\big) \sin\big((a - b)/2\big) \qquad (6.9e)$$

we obtain

$$P(\vartheta) = P_0 \Delta f + \{2/(2\pi D \sin \vartheta)/c\} \sin\left(2\pi \frac{D}{c} \frac{\Delta f}{2} \sin \vartheta\right) \cos\left(2\pi \frac{D}{c} f_0 \sin \vartheta\right) =$$

$$= P_0 \Delta f \left(1 + \frac{\sin\left(2\pi \frac{D}{c} \frac{\Delta f}{2} \sin \vartheta\right)}{\left(2\pi \frac{D}{c} \frac{\Delta f}{2} \sin \vartheta\right)}\right) \cos\left(2\pi \frac{D}{c} f_0 \sin \vartheta\right) =$$

$$= P_0 \Delta f (1 + si(x) \cos\{(2D f_0 \sin \vartheta)/c\} \qquad (6.9f)$$

$$x = 2\pi \frac{D}{c} \frac{\Delta f}{2} \sin \vartheta . \qquad (6.9g)$$

From this result it is obvious that in the case of a realistic frequency band instead of a single frequency the amplitudes of the power beam are additionally modulated by a sine function. If we again assume that for small angles $\sin \vartheta \approx \vartheta$ is valid, using equation (6.9f) we find that the amplitudes tend to zero when the *order of the interference* fringes m reaches the value $f_0/\Delta f$, since

$$si\left(2\pi \frac{D}{c} \frac{\Delta f}{2} \sin \vartheta_{\max}\right) = si\left(2\pi \frac{D}{c} \frac{\Delta f}{2} \vartheta_{\max}\right)$$

where

$$\vartheta_{\max} = mc/(f_0 D) \text{ by (6.8)}$$

$$= si\left(2\pi \frac{D}{c} \frac{\Delta f}{2} \frac{cf_0}{D\Delta f . f_0}\right)$$

$$si(\pi) = 0 \qquad (6.9h)$$

on monochromatic radiation therefore causes an attenuation of the interference amplitudes. In order to keep this effect negligible up to m-th order the receiver bandwidth Δf must be limited so that $\Delta f \ll f_0/m$. If a *broad bandwidth* is used then the high-order interference lobes are smeared out and the interferometer has a single central lobe. This effect was first utilized by Vikevich (SU) in 1953 (91) to construct a broadband short-wave radio interferometer UTR 2 (Ukrainian T-Radiotelescope, 1860 × 60.m) near Khar kow/SU which had one central lobe with a high angular resolution. It operated in the RF range 10. < RF < 25. MHz. UTR1 was a smaller prototype instrument of UTR2.

6.3. Effect of Extension of the Radiations Source on the Power Pattern

If the radio source has an extension in two dimensions the effects on the interferometer beam must be investigated (7,15,36). The source is assumed to have an angular extent w, within which the brightness distribution is given by $T_s(\vartheta)$. The source moves slowly through the pattern of lobes within the interferometer beam. If the source is assumed to be composed of a number of large number of point sources then the resulting interference pattern is a superposition of the patterns from the individual sources. We again assume that w is small so that $\sin \vartheta \approx \vartheta$. Using equation (6.6) and convolving $P(\vartheta)$ we obtain the power brightness distribution $T_s(\vartheta)$ and the power at the antenna output, setting the convolution to zero outside the integration limits.

$$P(\vartheta, D/\lambda) = \int_{-w/2}^{+w/2} T(\vartheta').\left\{1 + \cos\left((2\pi D/\lambda)(\vartheta + \vartheta')\right)\right\}d\vartheta' =$$

$$= \int_{-w/2}^{+w/2} T(\vartheta')\left\{1 + \cos\left((2\pi D\vartheta/\lambda).\cos(2\pi D\vartheta'/\lambda) - \right.\right.$$

$$\left.\left. - \sin(2\pi D\vartheta/\lambda).\sin(2\pi D\vartheta'/\lambda)\right)\right\}d\vartheta' =$$

$$= \int_{-w/2}^{+w/2} T(\vartheta')d\vartheta' + \int_{-w/2}^{+w/2} T(\vartheta')\cos(2\pi D\vartheta'/\lambda).\cos(2\pi D\vartheta/\lambda)d\vartheta' -$$

$$- \int_{-w/2}^{+w/2} T(\vartheta')\sin(2\pi D\vartheta'/\lambda)\sin(2\pi D\vartheta/\lambda)d\vartheta' . \tag{6.10}$$

Substituting

$$V \cos \alpha = \int_{-w/2}^{+w/2} T(\vartheta')\cos(2\pi D\vartheta'/\lambda)d\vartheta' \tag{6.11a}$$

$$V \sin \alpha = \int_{-w/2}^{+w/2} T(\vartheta')\sin(2\pi D\vartheta'/\lambda)d\vartheta' \tag{6.11b}$$

we obtain

$$P(\vartheta, D/\lambda) = \int_{-w/2}^{+w/2} T(\vartheta')d\vartheta' + V.\cos\{(2\pi D\vartheta/\lambda) - \alpha\} . \tag{6.12}$$

The output function of the interferometer therefore contains two components: a constant term corresponding to power received by each interferometer element and a second term which varies sinusoidally. If equation (6.11) is rewritten using complex notation it can be seen that the amplitude and phase of these terms are related to amplitude and phase of the brightness distribution of the source (15)

$$V \exp(j\alpha) = \int\limits_{-w/2}^{+w/2} T(\vartheta') \exp(j2\pi D\vartheta'/\lambda) d\vartheta' \ . \tag{6.13}$$

This term corresponds to a Fourier component of the brightness distribution which can be measured at the spatial frequency D/λ. If these measurements are extended over all interferometer baseline lengths D, the complete *Fourier spectrum* of the brightness distribution may be obtained.

The amplitude of the sinusoidally varying term is a function of the baseline length expressed in wavelengths and is frequently given in normalized form

$$0 \leq V_n(D/\lambda) = V(D/\lambda)/\big(V(0)\big) \leq 1 \ , \tag{6.14}$$

where

$$V(0) = \int T(\vartheta')d\vartheta' \ . \tag{6.15}$$

The parameter V_n is called the *visibility*. Following Michelson's nomenclature (92) an equivalent expression

$$V_n = (P_{\max} - P_{\min})/(P_{\max} + P_{\min}) \quad \text{and} \quad P = P(\vartheta, D/\lambda, w) \tag{6.16}$$

is also valid. The visibility is frequently given as a function of spatial frequency D/λ (4).

If the *brightness distribution* over a radio source of angular extent w is constant then it is possible to evaluate equation (6.10) directly

$$P(\vartheta, D/\lambda, w) = \int\limits_{-w/2}^{+w/2} T(\vartheta') \left\{ 1 + \cos(2\pi D(\vartheta + \vartheta')/\lambda \right\} d\vartheta' \ =$$

$$= \int\limits_{-w/2}^{+w/2} T(\vartheta')d\vartheta' + T(\vartheta') \left\{ \frac{\cos(2\pi D\vartheta/\lambda)}{2\pi D/\lambda} \ \sin(2\pi D\vartheta'/\lambda) - \right.$$

$$\left. - \frac{\sin(2\pi D/\vartheta/\lambda)}{2\pi D/\lambda} \ \cos(2\pi D\vartheta'/\lambda) \right\} \ =$$

$$= T(\vartheta')w + \left\{ T(\vartheta')/(2\pi D/\lambda) \right\}\Big|_{-w/2}^{+w/2} \ .$$

$$. \left\{ \big(\cos(2\pi D\vartheta/\lambda) \big) . \big(\sin(2\pi D w/2\lambda) \big) - \sin\big(2\pi D(-w)/(2\lambda) \big) - \right.$$

$$\left. - \big(\sin(2\pi D\vartheta/\lambda) \big) . \big(\cos(2\pi D w/2\lambda) \big) - \cos\big(2\pi D(-w)/(2\lambda) \big) \right\} \ =$$

$$= T(\vartheta')w + T(\vartheta')/(2\pi D/\lambda).$$

$$. \left\{ \big(\cos(2\pi D\vartheta/\lambda) \big) . \big(+2 \cos\big\{ (1/2) \big(2\pi \frac{Dw}{2\lambda} - 2\pi \frac{Dw}{2\lambda} \big) \big\} . \right.$$

$$. \big(\sin \big\{ (1/2) \big(2\pi \frac{Dw}{2\lambda} - 2\pi \frac{Dw}{2\lambda} \big) \big\} - \big(\sin(2\pi D\vartheta/\lambda) \big) .$$

$$\left. . \big(-2 \sin \big\{ (1/2) \big(2\pi \frac{Dw}{2\lambda} - 2\pi \frac{Dw}{2\lambda} \big) \big\} . \big(\sin \big\{ (1/2) \big(2\pi \frac{Dw}{2\lambda} + 2\pi \frac{Dw}{2\lambda} \big) \big\} \right\} \ .$$

$$(6.17a)$$

or

$$P(\vartheta, D/\lambda, w) = T(\vartheta')w\left\{1 + \left(\left(\sin(2\pi Dw/2\lambda)\right)/(2\pi Dw/2\lambda)\right)\cos(2\pi D\vartheta/\lambda)\right\} =$$
$$= T(\vartheta')w\{1 + si(x)\cos(2\pi D\vartheta/\lambda)\} \qquad (6.17b)$$

where

$$x = 2\pi Dw/2\lambda .$$

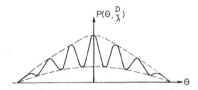

Fig. 6.4. Power pattern for an extended source (35).

Fig. 6.4 gives the power pattern for the case of a source with appreciable angular extent. Inspection of equation (6.16) shows that the function V_n has a maximum for $P_{\min} = 0$ as one expects for two equal monochromatic sources, and that V_n becomes zero if $P_{\min} = P_{\max}$ the interference lobes disappear. This we can write

$$P_{\max} = T(\vartheta')w\{1 + si(x)\} \qquad (6.18a)$$

and

$$P_{\min} = T(\vartheta')w\{1 - si(x)\} . \qquad (6.18b)$$

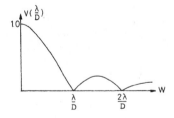

Fig. 6.5. Visibility function (4,24).

The visibility function can thus be written as

$$V_n(D/\lambda) = \frac{T(\vartheta')w\left\{\left(1 + si(x)\right) - \left(1 - si(x)\right)\right\}}{T(\vartheta')w\left\{\left(1 + si(x)\right) + \left(1 - si(x)\right)\right\}} = si(x) \qquad (6.18c)$$

or

$$V_n(D/\lambda) = \sin\{2\pi(D/\lambda)(w/2)\}/\{2\pi(D/\lambda)(w/2)\} . \qquad (6.18d)$$

The shape of this function is given in Fig. 6.5. $V_n(D/\lambda)$ decreases with increasing size w of the source and tends to zero at $w = \lambda/D$.

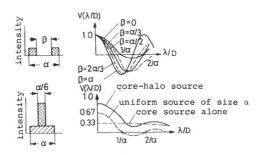

Fig. 6.6 Source models with corresponding λ/D variation (246).

By varying λ/D, i.e. by changing D for constant λ (65) it is possible to determine the angular extent of the source and the amplitude distribution within this angular extent. It is also possible to vary λ while keeping D constant but this is only possible within limits since the sizes of radio sources are frequency-dependent.

It is often the case that even approximate information about the size and shape of a radio source can be very useful to astronomers, for example whether the source is of core-halo structure. Fig. 6.6 shows the visibility as a function of λ/D which can be derived by using equation (6.13) with the intensity distribution for the source. The cases shown are a ring source and a pillbox structure or halo with a bright compact core.

The theory discussed here was first put to practical use in a sea interferometer by McCready (44) in Australia in 1946 (see Fig. 7.1).

7. RADIOASTRONOMICAL INTERFEROMETERS

7.1. Sea Interferometer

As described above, an interferometer requires two similar antenna systems, which are separated by some distance (6,15). The first interferometer constructed for radioastronomy, the sea interferometer of McCready, Pawsey and Scott at Dover Heights near Sydney (8,44) made use of a single antenna, since this instrument made use of a principle familiar in the field of optics. The single antenna had a horizontal beam, and was situated on a high cliff overlooking the sea. A wave packet arriving with a small angle to the horizontal would then arrive at the antenna in two ways, both directly and after reflection by the sea, which is a good reflector for short radio waves at low angles of incidence. This reflected wave packet has, when it reaches the antenna situated 80 metres above the sea, the same time difference relative to direct reception as it would have had if it had continued to an antenna at a height $h = -80$ metres. The effective interferometer baseline was thus $D = 160$.m. An antenna which is movable in azimuth allows observation of objects at various positions on the sky. Since the method only functions when the inclination of the arriving radiation to the horizontal is small, one is restricted to observations approximately an hour after rising and before setting of the celestial object. For such measurements at low elevations it is essential to apply corrections for atmospheric refraction (27). This instrument was used for solar measurements and for position and intensity measurements of radio sources (Fig. 7.1). This type of interferometer is analogous to the Lloyd's mirror interferometer (35,44).

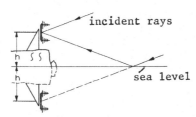

Fig. 7.1. Sea Interferometer
(Lloyd's mirror i (44)).

7.2. Interferometer with Passive Reflectors

If the costs of cables from the antenna to the central receiver are too high, or if there are problems connected with laying cables, one can use a radio link instead (44). A simpler technique was developed by the Leningrad group at Pulkovo (USSR) (33,51), using two passive mirrors to reflect the incoming waves towards the receiving antennas (Fig. 7.2).

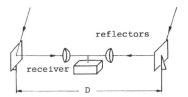

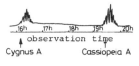

Fig. 7.2. Interferometer
with passive
reflectors (44).

Fig. 7.3. Recording made
with Ryle's
twin interferometer
(96,97,100).

7.3. Ryle's Twin Interferometer

The first interferometer to use two antenna elements was constructed by Ryle near Cambridge (UK) (28,45,97,189). Originally the antennas were in the form of two groups of 80 dipoles with a total length of 80.m. The two groups were spaced by 400.m and operated at 3.7m wavelength, the interference lobes were therefore 0.55° wide. The beam of each antenna group determines the overall beam of the interferometer and therefore the number and amplitude of the lobes In this case the width of the pattern was 5.2° so that approximately 9 lobes could be seen. It is advantageous that the number of lobes within the beam is small, since otherwise it is not possible to identify the central lobe and the advantages of having narrow lobes are lost. Fig. 7.3 shows a typical recording made with this system.

7.4. Phase-Switched Interferometer

Measurements of radio sources are always made in the presence of sky background noise against which these sources have to be distinguished. Receiver noise, atmospheric noise and ground radiation are also present (15,44).

If for example the position of a weak radio source seen against the strong background radiation of the milky way is to be determined, the ouput of the interferometer described above may be completely dominated by noise, as shown in Fig. 7.4a. In order to eliminate this problem Ryle changed the length of the cable to one antenna by one half wavelength at a frequency f_0, thus moving the whole interferometer pattern by one half lobe on the sky. The frequency f_0 is typically several hundred Hz so that the signal can be clearly distinguished from noise (Fig. 7.4b). The output function of a conventional interferometer is given by equation (6.6)

$$P(\vartheta, D/\lambda) = P_0(\vartheta)\{1 + \cos(2\pi D/\vartheta/\lambda)\} . \tag{7.1a}$$

All interfering signals are contained in the constant term of equation (6.6), thus the shift by half a wavelength leads to a different interferometer output function

$$P(\vartheta, D/\lambda) = P_0(\vartheta)\{1 - \cos(2\pi D\vartheta/\lambda)\} . \tag{7.1b}$$

A phase demodulator or phase-sensitive detector (PSD) (32) which is commutated at the switching frequency f_0 enables us to form the difference of the two states of the output function (Fig. 7.6).

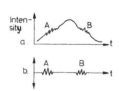

Fig. 7.4. *Principle of phase switching*
a. signal response of simple inter-
ferometers to discrete radio sources
in the presence of a strong back-
ground signal
b. signal response after elimina-
tion of the background noise by
phase switching.

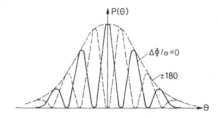

Fig. 7.5. Phase-switched power pattern.

In this way the constant term is cancelled and the resultant output function of such a *phase-switched* interferometer is

$$P(\vartheta, D/\lambda) = 2P_0(\vartheta) \cos(2\pi D\vartheta/\lambda) \ . \tag{7.2}$$

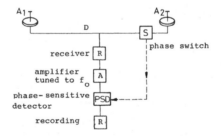

Fig. 7.6. Phase-switched twin interferometer.

7.5. Swept-Frequency Interferometer

The unwanted effects of background sources can also be eliminated by sweeping the radio frequency (19,44). The output function of an interferometer whose element antennas are connected by cables of unequal length to a swept-frequency receiver operating at a centre frequency f can be derived from expressions for frequency modulation (32)

$$P(\vartheta, D/\lambda) = P(\vartheta, Df/c) = P_0(\vartheta) \cos\{2\pi f(D \sin \vartheta + l)/c\} \ , \tag{7.3}$$

where l is the difference in cable length. The radio source lies in the direction ϑ and has a broad frequency spectrum. If now the receiver is tuned in rapid sequence to the radio frequencies f_1 and f_2, then the output function of the receiver shows oscillating interference fringes (Fig. 7.7).

Fig. 7.7.   Output function of a
swept-frequency interferometer.

The amplitude maxima of these interference fringes lie at

$$D \sin \vartheta + l = n.c/f \tag{7.4a}$$

and the amplitude minima at

$$D \sin \vartheta + l = \{n + (1/2)\}.c/f , \tag{7.4b}$$

where n is an integer. The fringe spacing, i.e. the frequency interval between successive peaks is given by

$$\Delta f = c/(D \sin \vartheta + l) . \tag{7.5}$$

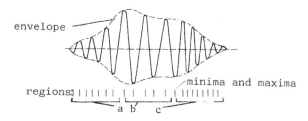

Fig. 7.8.   Interferogram of a radio flare on the sun observed
with a swept-frequency interferometer

region a: The source is fixed relative to the sun's surface. The
period of the interference fringes is that obtained for
the centre of the sun.

region b: The flare moves perpendicular to the sun's apparent
motion. Its angular velocity relative to the observer
is very small.

region c: The source follows the sun's apparent motion. The
resulting angular velocity relative to the observer is
very high and an increased fringe rate is seen (44).

From a measurement of the fringe spacing Δf relation (7.5) allows to find the position of a radio source at a certain fixed time. This method was developed by Wild and Sheridan/Australia in 1958 (101,262) in order to locate rapidly varying radio flares on the sun (15).

It is inherent to interferometer operation that the *period* of the interference fringe is determined by the *angular velocity* of the source. In the case of radiating sources on the sun this angular velocity has two components, the apparent movement of the sun itself and the movement of the flares over the sun. The flares move rapidly and are of such short duration that they cannot be located by a simple twin interferometer. The swept-frequency interferometer offers the advantage that the number of interference fringes received per second can be increased artificially by sweeping the radio frequency (Fig. 7.8).

7.6. Lobe Sweeping Interferometer

Many interferometers used in radio astronomy are transit instruments with which observations of a given object can be repeated once per day, the measurement time available is the time for the source to cross the antenna beam. Such an instrument cannot track rapid source variations.

The time the source requires to pass through the beam is determined by the baseline D, the *declination* δ of the source and the angular velocity $\dot{\rho}$ of the Earth. The recording time, i.e. the time to pass through one lobe is therefore

$$T = (D/\lambda)/(\dot{\rho} \cos \delta)(\text{seconds}) \qquad (7.6)$$

where $\dot{\rho} = 2\pi/(\text{rad}/24h) = 7.27\,10^{-5}$ radians/sec. For example a source at declination zero passes through the lobes of a meridian transit interferometer with baseline 100λ in 138 seconds. One cannot measure rapidly varying sources such as solar flares when, as in this case, the fringe period amounts to several minutes, but on the other hand use of longer baselines such as 1000λ leads to rapid fringe rates and the need to reduce the receiver time constant, reducing sensitivity. In order to avoid these difficulties the lobe sweeping interferometer was developed. The direction of the maximum of the reception pattern can be moved electrically by inserting a phase-shifter in the connection to one antenna. Thus the beam can for instance be made to sweep over the whole surface of the sun in a fraction of a second in order to locate rapid flaring sources. This technique was first used by Little and Payne-Scott in Australia in 1960 (90).

7.7. Multi-Element or Grating Interferometer

An antenna with a sharp pencil beam would be the ideal instrument to investigate large radio sources such as the sun, but single-dish antennas generally do not provide the required resolution. In principle one could use a two-element interferometer, carrying out a sequence of measurements at different baselines and therefore building up a picture of the source over a period of time. This method only functions if the source remains unchanged during the time (up to several months) that this procedure requires, but is often employed to obtain high resolution of radio sources. Baselines of over 60000λ leading to a resolution of a few arc seconds are typically employed.

To obtain narrow beams together with high time resolution the *multi-element interferometer* is used (15,44,90).

We consider the case of three identical antennas A_1, A_2, and A_3 situated on an East-West baseline as in Fig. 7.9, connected to a receiver by equal cables. A_1 and A_2 form an interferometer of baseline $s_0\lambda$ responsive to the spatial frequency s_0, i.e. features of this angular frequency must exist in the observed brightness distribution.

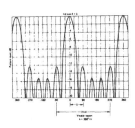

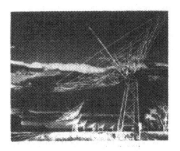

*Linear six-element array
grating lobe(s);
uniform distribution.*

*Circular 96-element Culgoora/
Australia array:
radio heliograph*
$D_{ar} = 3.$ *km*
$\lambda = 3.75$ *m*
(19, p. 67).

The interferometer A_2A_3 is similar to A_1A_2, thus A_1A_3 has a baseline of $2s_0$ and responds to a spatial frequency of $2s_0$. The angular spectrum of the system is shown in Fig. 7.10.

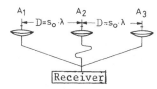

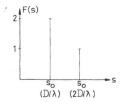

Fig. 7.9.   *Three-element
interferometer.*

Fig. 7.10.   *Spatial-frequency spectrum
of a three-element
interferometer.*

A *grating interferometer* with only three elements is in fact seldom used in practice, instead arrays of typically 8, 16 or 32 antenna elements spaced by $s_0\lambda$ are used. An instrument with 32 antennas has an amplitude for a spatial frequency s_0 31 times bigger than that of a two-element interferometer, for $2s_0$ the amplitude is 30 times longer, etc. This amplitude spectrum is shown in Fig. 7.12.

It can be seen that it is similar to that of a single filled aperture. The spectral lines are located at

$$s = (n-1)s_0 = (n-1).D/\lambda = f_0, 2f_0, \ldots , \tag{7.7a}$$

where n is the number of elements. $31s_0$ is the total baseline of the array expressed in wavelengths and also the highest spatial frequency to which this instrument is sensitive. A *multi-element* interferometer behaves as a *low pass filter*. For the computation of the reception pattern it is sufficient to add all signal amplitudes received while observing a radio source. The dashed line in Fig. 7.11a shows the power received by the antenna pair A_1A_2 as a function of ϑ. As the ϑ-dependence of the pair A_2A_3 is identical to that of A_1A_2 the behaviour of the combination of both pairs can be obtained by simple addition. The pattern of A_1A_3 has in fact the same amplitude as that of the pairs A_1A_2 or A_2A_3 but the angular period is $s_0/2$, since the distance between antennas is $2s_0$ (Fig. 7.11b).

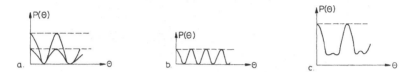

Fig. 7.11.   Power pattern of a three-element grating interferometer
a. elements A_1, A_2
b. elements A_1, A_3
c. elements A_1, A_2, A_3 (superposition of a. and b.).

By addition the pattern of the complete multi-element interferometer has the form of Fig. 7.11c. In the cases of 16 or 32 elements similar patterns of greater complexity are obtained. The ϑ-pattern of such a multi-element (n) arrangement is of form familiar for broadside arrays (23,50)

$$P = P_0(\vartheta) \, \frac{\sin^2\{(n\pi D \, \sin \vartheta)/\lambda\}}{\sin^2\{(\pi D \, \sin \vartheta)/\lambda\}} \; . \tag{7.7b}$$

If the baseline B between the elements is less than a wavelength, then equation (7.7b) is identical to that of a *broadside array* (36, p. 151). If D increases to more than one wavelength then the array pattern consists of a series of narrow lobes of *half power beam width* (HPBW) of about $\lambda/(n.D)$. The envelope of this pattern corresponds again to the pattern for a single element (see Fig. 7.12).

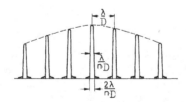

Fig. 7.12. Array factor of a grating interferometer.

If the total length of the grating interferometer is kept constant and the number of elements is doubled, then the number of lobes will be halved and the *angular* spacing will be doubled. The *sensitivity* of such an array is not confined to a finite number of *discrete* spatial frequencies but extends over a *continuous spectrum* of spatial frequencies whose upper bound is fixed by the total length of the array $D_t = (n - 1)s_0$. A very extended cylindrical parabolic reflector as used in Pulkovo (Fig. 7.14) can be regarded as a multi-element interferometer which consists of an infinite number of elements. The analogy between this system and an *optical diffraction grating* is another reason for treating this system as an interferometer.

Christiansen and Warburton built the first grating interferometer which consisted of 32 parabolic reflectors of $D_p = 2$.m diameter $\Delta D = 7$.m near Sydney. Operating at $\lambda = 21$.cm, it had a HPBW = 3.′and an *angular period* of 1.7°. This instrument was a heliograph, i.e. specially designed for solar observations. Since the angular extent of the sun is only $\Delta \varphi = 0.5°$ the repetition of the lobes is of no consequence. This equipment was able to measure accurate positions of solar flares. Later this instrument was supplemented by an array of 16 elements arranged at a right angle. Another 32 element grating interferometer is located near Nancay in France (44,115).

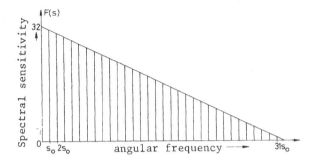

Fig. 7.13. *Spatial frequency spectrum of a 32 element grating interferometer.*

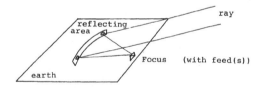

Fig. 7.14. *Parabolic reflector.*

7.8. Christiansen-Cross and Mills-Cross Interferometer

The systems considered so far were constructed on an east-west baseline and thus had a low resolution in the north-south direction. This is often a disadvantage, particularly where it is required to discriminate between sources which are close together in declination (85,107).

Normally the distance between two sources is determined by taking advantage of the earth's rotation. The angle of the connecting line between the sources and the west direction is a function of time. From multiple measurements of the direction of this connecting line or *parallactic angle* (68) it is possible to measure the source separation. In order to measure some separation in any direction in a single measurement one uses *systhesis* to stimulate a grating array with equal angular resolution in all directions.

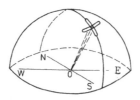

Fig. 7.15. Principle of the Mill-cross array.

This may be achieved by the use of two extended arrays of dipoles, one in the east-west and the other in the north-south direction, which intersect at the middle. The radiation pattern of this system projects a cross-shaped pattern on the celestial sphere (Fig. 7.15), since the arrays have orthogonal fan beams of the same polarization. In order to obtain a sharp single beam with negligible sidelobes one has recourse to a trick. The two antenna arrays are connected to the receiver by separate cables and added there in phase. The system receives signals from the whole cross-shaped beam, double the power being received where the fan beams intersect. The output voltage of the whole array is therefore

$$\underline{U}(\vartheta,\varphi) = \underline{U}_1(\vartheta,\varphi) + \underline{U}_2(\vartheta,\varphi) \; , \tag{7.8}$$

where $\underline{U}_1(\vartheta,\varphi)$ and $\underline{U}_2(\vartheta,\varphi)$ represent the voltages that are induced in each of the arrays by a radio source in the direction (ϑ,φ). Therefore the output power of the system is

$$P_+(\vartheta,\varphi) \propto \{\underline{U}_1(\vartheta,\varphi) + \underline{U}_2(\vartheta,\varphi)\} \{\underline{U}_1(\vartheta,\varphi) + \underline{U}_2(\vartheta,\varphi)\}^* \; . \tag{7.9}$$

If both array systems are operated in *anti-phase*, which can for instance be achieved by lengthening one transmission line by half a wavelength, then again radiation will be received in a cross-shaped pattern, but signals received in the intersection region will mutually cancel (Fig. 7.17),

Fig. 7.16. In-phase characteristics of the Mills-cross.

Fig. 7.17. Anti-phase characteristics of the Mills-cross.

and the array output voltage is

$$\underline{U}(\vartheta,\varphi) = \underline{U}_1(\vartheta,\varphi) - \underline{U}_2(\vartheta,\varphi) \; . \tag{7.10}$$

The power is

$$P_-(\vartheta,\varphi) \propto \{\underline{U}_1(\vartheta,\varphi) - \underline{U}_2(\vartheta,\varphi)\} \{\underline{U}_1(\vartheta,\varphi) - \underline{U}_2(\vartheta,\varphi)\}^* \; . \tag{7.11}$$

If we now switch the array between in-phase and anti-phase at a fixed frequency, the radiation received from a source with the highly directive central lobe of the cross-shaped pattern will be modulated, while the radiation from a source outside this region remains unaffected by a switching. The modulated output function of this array is proportional to the difference of both possible output functions and is given by

$$P = P_+ - P_- \propto \underline{U}_1(\vartheta,\varphi)\underline{U}_2^*(\vartheta,\varphi) + \underline{U}_1^*(\vartheta,\varphi)\underline{U}_2(\vartheta,\varphi) \ . \tag{7.12}$$

This output is processed further in a narrow-band amplifier and a *phase-sensitive detector* (32), so that finally only the energy in the central lobe is recorded. By varying the phase relationships the highly directive lobe of the cross array can be swept in the *meridian* plane in order to be able to observe different objects at zenith distances up to 45° (18,21).

The first large cross array was built in 1953 by Mills at Fleurs in Australia, and for this reason the system is referred to as the *half-wave dipoles* Mills-cross (103–108,110–113).

Each of the arms consisted of 500.*half-wave dipoles* with a total length of 460.m and an angular resolution of $\vartheta_{-3} = 0.4°$ at $\lambda = 3.5m$ wavelength. In 1964 a new Mills-cross (107,108) was erected near Camberra with an *arm length* of one nautical mile.

It should be mentioned that—as for every array—the equivalent collecting area and the *directivity* of the Mills-cross is far lower than for an equivalent, completely filled square aperture and equal to twice the geometrical mean of the aperture of each single arm. The main advantage of this system lies in the limitation of the number of antenna elements, reducing the cost of the system. The sidelobe level of the original Mills-cross was 13.*dB* (6,15).

Fig. 7.18.   Principle of the Christiansen-cross interferometer.

In order to produce a system with multiple *pencil beams* one combines the principle of the Mills-cross with that of a grating lobe interferometer. This type of interferometer consists of two multi-element interferometers which again intersect at a right angle (Fig. 7.18). Each grating interferometer has a special array pattern which consists of a series of fan beams. If the output functions of both systems are multiplied and time-averaged the resulting power pattern is proportional to the product of the voltage response of the interferometer systems. By increasing the angular distance between these lobes so that only a single lobe sees the radio source, the source may be probed by sweeping a rapid sequence of lobes across the sky using phase shifters. Thus a two-dimensional image of the source can be formed. The first multi-element cross interferometer was introduced in 1958 by Christiansen and Mathewson (110,103,104,105,106,111,112,113,6,7). The instrument was used to obtain images of the sun at 1.42 GHz. The interferometer consisted of two multi-element interferometers, each with 32 parabolic reflectors of 6.m diameter spaced by 33.m.

A similar instrument operating at a frequency of $f = 3.3$ GHz was erected in 1961 under the leadership of Bracewell and Swarup at Stanford (USA) (2,3,24) consisting of two 16-element interferometers of $D = 125.m$ (1255λ) each. The diameter of the parabolic reflector elements was $D_p = 3.m$ and the resolution $\Delta\vartheta_3 = 3.'$. With the steady improvement of the angular resolution of radio telescopes also the number of sources to be investigated increased. As a great deal of time would be required to survey an extended region by sequential sampling it becomes more reasonable to simultaneously sweep with all available lobes over the sky. In this procedure the output signal of each interferometer element is first amplified and then divided into m different output signals which are applied to m different receivers of identical phase performance. The m outputs of the amplifiers correspond to m different narrow lobes which point to m points on the sky. A simultaneous multi-lobe interferometer was developed and erected in 1961 by Blum (115,44) for Meudon Observatory (France). This sytem used two linear arrays, aligned east-west and north-south. The east-west array, consisting of 32 parabolic reflectors, had a total length of $D_t = 1550.m$, the north-south array was 700.m long and used eight reflectors. The HPBW of the central lobe at 167.MHz was 3.4' and 7' in east-west and north-south directions respectively. Using the scheme illustrated in Fig. 7.19 15 pencil beam lobes could be obtained. Positions of radio sources could be measured with an accuracy of 0.5 arc minutes in declination (24,85) and 0.25 arc minutes in right ascenscion.

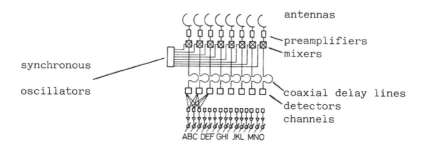

Fig. 7.19.   Block diagram of an interferometer with 15 simultaneous pencil beams.

7.9. Correlation Systems

A measure of the coherence between two periodic signals which are shifted in time by τ is the *crosscorrelation*, which can be visualized as the average over a long period of time of the product of two signals (63)

$$r(\tau) = \int\limits_{-\infty}^{+\infty} s_1(t)s_2(t-\tau)dt \ . \tag{7.13}$$

The expression

$$\rho(t) = \left\{ \int\limits_{-\infty}^{+\infty} s_1(t)s_2(t-\tau)dt \right\} / \left\{ \int\limits_{-\infty}^{+\infty} s_1^2(t)dt \int\limits_{-\infty}^{+\infty} s_2^2(t)dt \right\}^{1/2} , \tag{7.14a}$$

is referred to as the *crosscorrelation coefficient* (CCC), which using the *Schwarz inequality (8.85)*, becomes unity for complete identity of the signals to be compared.

If we have the case (77)

$$s_1(t) = s_2(t) \tag{7.14b}$$

then because

$$\int\limits_{-\infty}^{+\infty} s_1^2(t-\tau)dt = \int\limits_{-\infty}^{+\infty} s_2^2(t)dt \tag{7.14c}$$

we obtain

$$\rho(t) = \left\{ \int\limits_{-\infty}^{+\infty} s_1(t)s_1(t-\tau)dt \right\} \bigg/ \int\limits_{-\infty}^{+\infty} s_1^2(t)dt \ , \tag{7.15}$$

which is called the *autocorrelation coefficient* (acc). The acc (63) describes the similarity of the signal $s_1(t)$ to a copy of itself shifted in time by τ.

As the acc is only influenced by the time dependence of the signal and not by amplitude variations, only the numerator in equation (7.15)

$$r(\tau) = \int\limits_{-\infty}^{+\infty} s_1(t)s_2(t-\tau)dt \ , \tag{7.16}$$

influence the *autocorrelation function* (acf). For $\tau = 0$ the acc becomes one as the auto-correlation *function* acf tends to

$$r(0) = \int\limits_{-\infty}^{+\infty} s_1^2(t) = E \ , \tag{7.17}$$

where E represents the finite energy of the signal.

Theoretically, however, events exist for which the energy E is infinite, since they do not cease after a certain time. For a certain class of these events, for example stationary random thermal noise of a resistor at constant temperature, there exists a limiting value

$$\overline{s^2(t)} = \lim_{T \to \infty} \frac{1}{2T} \int\limits_{-\infty}^{+\infty} s^2(t)dt \ . \tag{7.18}$$

If, e.g. $s(t)$ is a voltage then equation (7.18) represents power produced by a resistor. Since a process must be finite in order that a Fourier transform can exist then following Wiener (63) we consider only a short section of the process $s(t)$ instead of waiting for it to end.

$$\underline{s}_T(t) = \begin{cases} s(t) \text{ in the interval } -T < t < +T \\ 0 \text{ outside this interval} \ . \end{cases} \tag{7.19}$$

It is now possible to write the Fourier transform

$$\underline{S}(f) = \int\limits_{-\infty}^{+\infty} \underline{s}_T(t)\exp(-j2\pi ft)dt \ =$$

$$= \int\limits_{-T}^{+T} \underline{s}(t)\exp(-2j\pi ft)dt \ . \tag{7.20}$$

In order to determine the power we can write for finite T

$$\int\limits_{-T}^{+T} s^2(t)dt = \int\limits_{-\infty}^{+\infty} s_T^2(t)dt = \int\limits_{-\infty}^{+\infty} \underline{s}_T(t)s_T^*(t)dt \ . \tag{7.21}$$

Hence

$$\underline{s}_T(t) = \int\limits_{-\infty}^{+\infty} \underline{S}(f)\exp(+j2\pi ft)dt \tag{7.22a}$$

or

$$\underline{s}_T^*(t) = \int\limits_{-\infty}^{+\infty} \underline{S}^*(f)\exp(-j2\pi ft)df \tag{7.22b}$$

$$\int\limits_{-T}^{+T} s^2(t)dt = \int\limits_{-\infty}^{+\infty} \underline{s}_T(t)dt \int\limits_{-\infty}^{+\infty} \underline{S}_T^*(f)\exp(-j2\pi ft)df =$$

$$= \int\limits_{-\infty}^{+\infty} \underline{S}_T^*(f)df \int\limits_{-\infty}^{+\infty} \underline{s}_T(t)\exp(-j2\pi ft)dt =$$

$$= \int\limits_{-\infty}^{+\infty} \underline{s}_T(t)\underline{s}_T^*(t)df = 2\int\limits_{-\infty}^{+\infty} |\underline{S}_T(f)|^2 df \ . \tag{7.22c}$$

The power is then the *time average* of equation (7.22c)

$$\overline{s^2(t)} = \lim_{T\to\infty} \frac{1}{2T} \int\limits_{-\infty}^{+\infty} |\underline{S}_T(f)|^2 df = \lim_{T\to\infty} \frac{1}{T} \int\limits_{0}^{+\infty} |\underline{S}_T(f)|^2 df =$$

$$= \int\limits_{0}^{+\infty} X(f)df = (1/2)\int\limits_{-\infty}^{+\infty} X(f)df = N_S \ , \tag{7.23}$$

where

$$\lim_{T\to\infty} (1/T)|S_T(t)|^2 = X(f) \tag{7.24a}$$

is referred to a *spectral power density*. If we require the acf for a process of infinite length we obtain the following expression

$$r(\tau) = \lim_{T\to\infty} (1/2T) \int\limits_{-\infty}^{+\infty} \underline{s}_T(t-\tau)dt = \lim_{T\to\infty} (1/2T) \int\limits_{-T}^{+T} \underline{s}_T(t)s_T(t-\tau)dt \ . \tag{7.24b}$$

This gives a measure for the autocorrelation of all events obeying (7.18). The acf $r(\tau)$ has its maximum value at $\tau = 0$, namely

$$r(0) = \lim_{T\to\infty} (1/2T) \int\limits_{-T}^{+T} s^2(t)dt \ . \tag{7.24c}$$

Similarly, the *autocorrelation coefficient* (acc) may be defined as the autocorrelation per unit power

$$\rho(\tau) = \left\{ \lim_{T\to\infty} \frac{1}{2T} \int\limits_{-T}^{+T} \underline{s}(t)\underline{s}(t-\tau)dt \right\} / \left\{ \lim_{T\to\infty} \frac{1}{2T} \int\limits_{-T}^{+T} s^2(t)dt \right\} \ , \tag{7.25a}$$

which has its maximum value at $\tau = 0$.

The acf ρ and the power density s^2 are related to one another by the Fourier transform. We now form the *inverse* Fourier transform (72) of the spectral power density

$$(1/2) \int_{-\infty}^{+\infty} X(f) \exp(j2\pi f\tau)df = \lim_{T\to\infty} (1/2T) \int_{-\infty}^{+\infty} \underline{S}_T(f)\underline{S}_T^*(f) \exp(j2\pi f\tau)df \ . \qquad (7.25b)$$

If the spectrum $\underline{S}^*(f)$ is expressed by $\underline{s}(t)$, the following relation is obtained

$$\lim_{T\to\infty} (1/2T) \int_{-\infty}^{+\infty} \underline{S}_T(f)df \int_{-\infty}^{+\infty} \underline{s}_T(t) \exp(j2\pi ft) \exp(2\pi f\tau)dt \ =$$

$$= \lim_{T\to\infty} (1/2T) \int_{-\infty}^{+\infty} \underline{s}_T(t)dt \int_{-\infty}^{+\infty} S_T(f) \exp\big(j2\pi f(t+\tau)\big)df \ =$$

$$= \lim_{T\to\infty} (1/2T) \int_{-\infty}^{+\infty} \underline{s}_T(t)s_T(t-\tau)dt = \lim_{T\to\infty} (1/2T) \int_{-T}^{+T} \underline{s}_T(f)\underline{s}(t-\tau)dt = r(\tau) \ . \qquad (7.26a)$$

Thus the familiar *Wiener theorem* (63) is obtained

$$r(\tau) = \lim_{T\to\infty} (1/2T) \int_{-T}^{+T} \underline{s}_T(t)\underline{s}(t-\tau)dt \ =$$

$$= \lim_{T\to\infty} (1/2T) \int_{-\infty}^{+\infty} \underline{S}_T(f)\underline{S}_T^*(f) \exp(j2\pi f\tau)df \ . \qquad (7.26b)$$

As $r(\tau)$, the acf, is a real function, equation (7.26a) becomes

$$r(\tau) = \mathrm{Re}\left\{(1/2) \int_{-\infty}^{+\infty} X(f) \exp(j2\pi f\tau)df\right\} = \int_{0}^{+\infty} X(f) \cos(2\pi f\tau)df \ , \qquad (7.26c)$$

and it can be seen that the acf $r(\tau)$ is an even function for all signals. Forming the inverse Fourier transform for equation (7.26c) we obtain

$$X(f)/2 = \left\{\mathrm{Re} \int_{-\infty}^{+\infty} \overline{r(\tau)} \exp(-j2\pi f\tau)d\tau\right\} \qquad (7.26d)$$

$$X(f) = 4 \int_{0}^{+\infty} r(\tau) \cos(2\pi f\tau)d\tau \ . \qquad (7.26e)$$

The advantage of the acf is that it allows us to describe the spectral behaviour of a random process. For all random signals with identical acf the shape of the power spectrum is identical (63).

This result is of immediate application to an interferometer shown schematically in Fig. 7.20 (44). The signals received by the antenna elements A_1 and A_2 are amplified and sent by cables to a multiplier followed by an integrator, a combination referred to as a *correlator*.

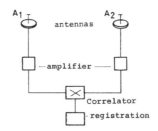

Fig. 7.20. Principle of a correlation interferometer.

The observed radio source induces *noise currents* $i_{1,2}(t)$ in the antenna elements, which are completely correlated as they come from one single source; here $i_2(t)$ is delayed due to the baseline by the time difference τ compared to the current $i_1(t)$ from A_1. The noise contributions from the amplifiers can be considered as completely independent of each other so that their product formed in the correlator is zero.

This scheme avoids the problem of separating the wanted signal from the white noise of the amplifiers.

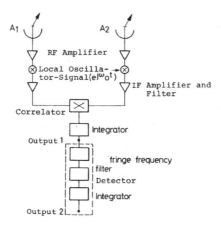

Fig. 7.21.   Correlation interferometer with interference fringe filter (254).

The first interferometer to make use of a correlation system of this type was developed by Blum in France in 1959 (115). Fig. 7.21 shows the block diagram of a correlation interferometer radio source inducing a voltage corresponding to a signal voltage $\underline{U}$ of

$$\mathrm{Re}\,\{\underline{U}(t)\exp(j(\omega_0 + \omega_1)t)\} \tag{7.27a}$$

within the required frequency band in the first antenna element A_1 and an equal voltage delayed by the time difference τ at the other element A_2 and if the receiver noise voltages are

$$\text{Re}\{\underline{N}_1(t)\exp(j(\omega_0+\omega_1)t)\} \tag{7.27b}$$

$$\text{Re}\{\underline{N}_2(t)\exp(j(\omega_0+\omega_1)t)\} , \tag{7.27c}$$

then the voltages at the correlator inputs are

$$\text{Re}\{\underline{U}(t)\exp(j\omega_1 t)+\underline{N}_1(t)\exp(j\omega_1 t)\} \tag{7.27d}$$

$$\text{Re}\{\underline{U}(t-\tau)\exp(j\omega_1 t)\exp(-j(\omega_0+\omega_1)\tau)+\underline{N}_2(t)\exp(j\omega_1 t)\} . \tag{7.27e}$$

The output of the correlator is then

$$\text{Re}\{\underline{U}(t)\underline{U}^*(t-\tau)\exp(j(\omega_0+\omega_1)\tau)+\underline{N}_1(t)\underline{U}^*(t-\tau)\exp(j(\omega_0+\omega_1)\tau +$$

$$+ \underline{U}(t)N_2^*(t)+\underline{N}_1(t)N_2^*(t)\} . \tag{7.27f}$$

If we assume that the delay τ is small compared to the correlation time t_c of $\underline{U}(t)$, so that $\underline{U}(t-\tau)\approx\underline{U}(t)$, and that the integrator is switched with the sampling frequency $f_s=1/\Delta t$ then the output function of the correlator is

$$\text{Re}\{\overline{\underline{U}(t)\underline{U}^*(t)}\exp(j(\omega_0+\omega_1)\tau)+\overline{\underline{N}_1(t)\underline{U}^*(t)}\exp(j(\omega_0+\omega_1)\tau) +$$

$$+ \overline{N_2^*(t)\underline{U}(t)}+\overline{\underline{N}_1(t)N_2^*(t)}\} , \tag{7.27g}$$

where the bar represents the time average over Δt.

The output function at output 1 is then

$$\underline{U}(t)\underline{U}^*(t)\cos(\omega_0+\omega_1)\tau . \tag{7.27h}$$

Therefore the square of the noise at the output is

$$N^2 = \left|\overline{\underline{U}(t)\underline{U}^*(t)}\exp\{(j(\omega_0+\omega_1)\tau)\}+\overline{\underline{N}_1(t)\underline{U}^*(t)}\exp j(\omega_0+\omega_1)\tau+\overline{N_2^*(t)\underline{U}(t)} +$$

$$+ \overline{N_1(t)N_2^*(t)^2}-\underline{U}(t)\underline{U}^*(t)\cos(\omega_0+\omega_1)\tau\right|^2 =$$

$$= \left|\{\overline{\underline{U}(t)\underline{U}^*(t)}-\underline{U}(t)\underline{U}^*(t)\}\cos(\omega_0+\omega_1)\tau+\overline{\underline{N}_1\underline{U}^*(t)}\exp\{j(\omega_0+\omega_1)\tau\} +$$

$$+ \overline{N_2^*(t)\underline{U}(t)}+\overline{\underline{N}_1(t)N_2^*(t)}\right|^2 =$$

$$= \left\{\left(\overline{\underline{U}(t)\underline{U}^*(t)}-\underline{U}(t)\underline{U}^*(t)\right)^2\cos^2(\omega_0+\omega_1)\tau+\left|\overline{\underline{N}_1(t)\underline{U}^*(t)}\right|^2 +$$

$$+ \left|\overline{N_2^*(t)\underline{U}(t)}\right|^2+\left|\overline{\underline{N}_1(t)N_2^*(t)}\right|^2\right\} . \tag{7.27i}$$

If we assume that $\underline{N}_1(t)$, $\underline{N}_2(t)$ and $\underline{U}(t)$ have the same power spectra, then we can make the following simplification using

$$s = \underline{U}(t)\underline{U}^*(t) \tag{7.28a}$$

$$n_1 = \underline{N}_1(t)\underline{N}_1(t) \tag{7.28b}$$

$$n_2 = \underline{N}_2(t)\underline{N}_2(t) \tag{7.28c}$$

$$N^2 = (1/\Delta t)C\{(1/2)s^2+n_1 s+n_2 s+n_1 n_2\} , \tag{7.29}$$

where C represents the integral of the square amplitude of the acf

$$C = \left|\frac{1}{\overline{U\,U^*}}\right| \int\limits_{-\infty}^{+\infty} |\underline{U}(t)\underline{U}(t-x)|^2 dx \;= \tag{7.30a}$$

$$= \left|\frac{1}{\overline{N_1 N_1^*}}\right| \int\limits_{-\infty}^{+\infty} |\underline{N}_1(t)\underline{N}_1(t-x)|^2 dx \;= \tag{7.30b}$$

$$= \left|\frac{1}{\overline{N_2 N_2^*}}\right| \int\limits_{-\infty}^{+\infty} |\underline{N}_2(t)\underline{N}_2(t-x)|^2 dx \;. \tag{7.30c}$$

The output function of a correlation interferometer of this kind is normally analyzed by *cross* correlating it with a reference sine wave of angular frequency $\omega_r = (\omega_0 + \omega_1)(d\tau/dt)$. For a period $m.\Delta\tau$ the amplitude error of the sine wave signal is $\sqrt{2/mN}$, which is assumed to be much smaller than s.

Another method shown by the area of Fig. 7.21 enclosed by dotted lines uses a *fringe filter* instead of an integrator. If this filter has a bandpass given by the integration time $m.\Delta t$ and the signal-to-noise ratio somewhat greater than one, then this method will lead to the same signal-to-noise ratio as the previous one, since the use of a fringe filter has the same effect as correlation with a sine wave.

Fig. 7.22a shows a typical *intensity* or *postdetection interferometer* (90,116,117), which will be discussed in the following section.

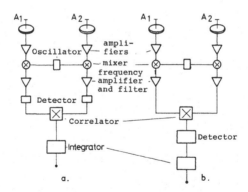

Fig. 7.22. *Postdetection correlation interferometer*
a. usual system
b. modified system.

The original system and a modified one are presented in Fig. 7.22a and b. Both versions are equivalent since considering voltages Re $\{\underline{U}_1\}$ and Re $\{\underline{U}_2\}$ at the outputs of the two intermediate frequency (IF) amplifiers then the output of the integrators of both systems show a signal of the form

$$s'^2 = \overline{\underline{U}_1\underline{U}_1^*\underline{U}_2\underline{U}_2^*} \;. \tag{7.31a}$$

The system of Fig. 7.22b corresponds to that of Fig. 7.21 except for the fringe filter. Consequently, a series of interferometer systems may be designed differing only in the bandwidth of this fringe filter. If the bandwidth is determined by total *integration time*, the system is a normal *correlation interferometer* except that the phase is not measured. If this filter has a width of double or more that of the intermediate frequency filter, this is referred to as an *intensity interferometer* (90,116).

We now calculate the signal-to-noise ratio of an interferometer intermediate between these types which has a fringe filter with limited bandwidth Δf_f. Assuming

$$\underline{U}(t)\underline{U}^*(t) = s \ll \underline{N}_1(t)\underline{N}_1^*(t) \tag{7.31b}$$

and

$$\underline{N}_1(t)\underline{N}_1^*(t) = \underline{N}_2(t)\underline{N}_2^*(t) = n \ . \tag{7.31c}$$

Then at the input of the fringe filter we have approximately

$$\text{Re}\,\{\underline{U}(t)\underline{U}^*(t)\exp(j(\omega_0 + \omega_1)\tau) + \underline{N}_1(t)\underline{N}_2^*(t)\}\ . \tag{7.32a}$$

At this point the power ratio of signal-to-noise is

$$(1/2)\,\{\underline{U}(t)\underline{U}^*(t)\}/\{\underline{N}_1(t)\underline{N}_1^*(t)\underline{N}_2(t)\underline{N}_2^*(t)\}\ . \tag{7.32b}$$

If the IF filters have a rectangular bandpass of bandwidth

$$\Delta f_{\text{IF}} \gg \Delta f_r \ \text{ and } \ f_{\text{IF}} \gg \omega_r/2\ , \tag{7.32c}$$

noise after the filter will be reduced by the ratio between the bandpass of the intermediate frequency filter and the bandpass of the fringe frequency filter. A simple expression for this band-limited noise yields the SNR at the output integrated over the time interval $\tau \ll 1/\Delta f_r$, assuming that the full *interferometer fringe power* passes through the fringe frequency filter (254)

$$S/N\ = (1/2a)\sqrt{\Delta f_f \tau}\,\{\underline{U}(t)\underline{U}^*(t)\}^2/\Big\{\Big(\underline{N}_1(t)\underline{N}_1^*(t)\underline{N}_2(t)\underline{N}_2^*(t)\Big)(\Delta f_f/\Delta f_{\text{IF}}) +$$
$$+ \Big(\underline{U}(t)\underline{U}^*(t)\Big)^2\Big\} = a\sqrt{\Delta f_f \tau}\,\Delta f_{\text{IF}} s^2/\Big\{2\Delta f_f.n^2\Big(1 + \Big(s^2\Delta f_{\text{IF}}/(n^2 f_f)\Big)\Big)\Big\}\ , \tag{7.32d}$$

where a is a constant of order unity. The further assumption is made when investigating the case of a high S/N that the interference fringe function has a bandwidth Δf_f, if this assumption is not valid for the term in the final brackets.

For small SNR the correction after the fringe frequency filter will be inversely proportional to the square root of Δf_p.

7.9.1. Hanbury Brown-Twiss or Intensity Interferometer (Post Detection Correlation Interferometer)

The angular resolution of a radio interferometer of Michelson type is *proportional* to its element spacing and therefore may be improved by increasing this baseline length. For a short baseline of only a few kilometers normally commercial *coaxial cables* are used to connect these elements to the receiver.

For longer baselines radio links are used (Fig. 7.2) which present difficulties in the maintainance of *phase stability* in the link and in the receiver equipment. Instabilities are caused by fluctuations in the propagation medium, i.e. atmospheric turbulence. Small phase instabilities may lead to shifts in the interference lobes causing them to be smeared.

If the baseline is extended beyond 50.km it becomes nearly impossible under normal atmospheric and ionospheric conditions to maintain phase stability. The baseline length therefore represents the limit of the conventional interferometer.

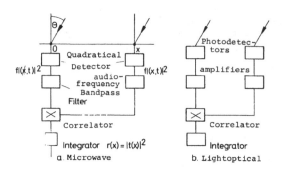

Fig. 7.23. The Hanbury-Brown-Twiss Interferometer.

Hanbury-Brown and Twiss (14) eliminated this difficulty by replacing the communication link to the distant elements by a more suitable arrangement. The block diagram of this system is given in Fig. 7.23 (14,90,116,117). The output of the antennas are applied to a square-law detector.

The low frequency signals at the detector output are sent to the correlator via audio links. The detector outputs represent intensity fluctuations which are not independent statistically.

After forming the time average the output of the intensity interferometer is a function of the form

$$r(x) = |\underline{f}_s(x)|^2 \tag{7.33}$$

where $\underline{f}_s$ corresponds to the spatial frequency of a brightness distribution $T(u)$. The relevant expressions are derived as follows (71,72,84,90,118):

The low frequency output functions of the detectors are $|f(0,t)|^2$ and $|f(x,t)|^2$. If these functions are sent to ideal bandpass filters which *eliminate* the DC terms but pass the variable components we obtain

$$|\underline{f}(0,t)|^2 - f_s(0) \quad \text{and} \quad |\underline{f}(x,t)|^2 - f_s(0) \,, \tag{7.34a}$$

where

$$f_s(0) = |\underline{f}(0,t)|^2 = |\underline{f}(x,t)|^2 \tag{7.34b}$$

corresponds to the DC component of the output functions. If these are combined in the correlator and time averaged we obtain

$$r(x) = \overline{\{|\underline{f}(0,t)|^2 - f_s(0)\} \{|\underline{f}(x,t)|^2 - f_s(0)\}} =$$
$$= \overline{|\underline{f}(0,t)|^2|f(x,t)|^2 - |\underline{f}(0,t)|^2 f_s(0) - |\underline{f}(x,t)|^2 f_s(0) + f_s(0)^2} \,. \tag{7.35}$$

Introducing the fourth-order moment (21)

$$\overline{\underline{f}_1 \underline{f}_2^* \underline{f}_3 \underline{f}_4^*} = \overline{\underline{f}_1 \underline{f}_2^* \, \underline{f}_3 \underline{f}_4^*} + \overline{\underline{f}_1 \underline{f}_4^* \, \underline{f}_2^* \underline{f}_3} \tag{7.36a}$$

and forming the following expressions

$$|\underline{f}(0,t)|^2 = \underline{f}(0,t)\underline{f}^*(0,t) \tag{7.36b}$$

$$|\underline{f}(x,t)|^2 = \underline{f}(x,t)\underline{f}^*(x,t) , \tag{7.36c}$$

then with

$$\underline{f}_s(x) = f(0,t)\underline{f}^*(x,t) \tag{7.36d}$$

equation (7.34) yields

$$
\begin{aligned}
r(x) &= \overline{\underline{f}(0,t)\underline{f}^*(0,t)\underline{f}(x,t)f^*(x,t) - f_s(0)\left\{\underline{f}(0,t)\underline{f}^*(0,t) + \right.} \\
&\quad \overline{\left. + \underline{f}(x,t)\underline{f}^*(x,t)\right\} + f_s(0)^2} = \\
&= \overline{\underline{f}(0,t)\underline{f}^*(0,t)}\,\overline{\underline{f}(x,t)\underline{f}^*(x,t)} + \overline{\underline{f}(0,t)\underline{f}^*(x,t)}\,\overline{\underline{f}^*(0,t)\underline{f}(x,t)} - \\
&\quad - f(0)\left\{\underline{f}(0,t)\underline{f}^*(0,t) + \underline{f}(x,t)\underline{f}^*(x,t)\right\} + f_s(0)^2 = \\
&= f_s(0)^2 + |\underline{f}_s(x)|^2 - f_s(0)\left\{f_s(0) + f_s(0)\right\} + f_s(0)^2 = \\
&= |\underline{f}_s(x)|^2 , \tag{7.37}
\end{aligned}
$$

which is the required result. If the baseline length is varied, the output function of this intensity interferometer is proportional to the squared absolute value of the spatial frequency spectrum of the source distribution. This distribution is generally complex, but since the argument of $\underline{f}_s(x)$ is eliminated by squaring *no phase* information may be obtained from the output signal of a normal intensity interferometer.

This is a disadvantage of the intensity interferometer which limits the amount of information it can give for a complex brightness distribution. The main advantage of the intensity interferometer is that it is applicable to very long baselines with resultant high resolution. It is also insensitive to phase fluctuations of the received signal. Its great disadvantage, however, is very low sensitivity compared with other systems so that it is only suitable for studies of very powerful radio sources.

The principle of the intensity interferometer has found practical application at optical wavelengths. The largest Michelson stellar interferometer at Mount Wilson Observatory has a 6.m baseline, attempts to further increase this figure failed owing to insurmountable technical problems. The optical intensity interferometer built by Hanbury-Brown and Twiss in 1958 to measure the diameter of Sirius A avoided such problems. The size of the star was determined to be 6.7×10^{-3} arcsec, too small to be resolved by a Michelson interferometer. The University of Sydney built a large stellar interferometer at Narrabri using two 30.m parabolic mirrors mounted on a circular railtrack of 200.m in diameter. A block diagram of such a system is shown in Fig. 7.23b (109,117).

7.9.2. Compound Intensity Interferometer

An ideal interferometer would combine the advantages of both the Michelson and Hanbury-Brown Twiss systems so that phase is conserved yet the RF link between distant antennas and the receiver are replaced by LF lines. By combining these two principles a compound intensity interferometer (33) was devised by Covington and Broten in 1957 (24,90).

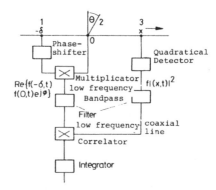

Fig. 7.24. The compound intensity interferometer.

Fig. 7.24 shows an array (2,3,4) of three elements, which are positioned at the points x, $x = 0$ and $x = -\delta$ on the horizontal axis. The signals received by the two local antennas 1 and 2 which have an arbitrary mutual phase Φ are combined as in the normal Michelson correlation interferometer in a multiplier. The signal from antenna 3 is square-law detected. Thus the two *low-frequency* output functions

$$\mathrm{Re}\,\{\underline{f}(-\delta,t)f^*(0,t)\exp(j\Phi)\} \quad \text{and} \quad |\underline{f}(x,t)|^2 \tag{7.39a}$$

are obtained.

As shown in Fig. 7.24 these signals are applied to a low-frequency bandpass filter in order to eliminate the DC components, the output is then

$$\mathrm{Re}\,\{\underline{f}(-\delta,t)\underline{f}^*(0,t)\exp(j\Phi)\} - \mathrm{Re}\,\{\underline{f}_s^*(\delta)\exp(j\Phi)\}\,, \tag{7.39b}$$

where the second term *corresponds to DC*

$$\mathrm{Re}\,\{\underline{f}_s^*(\delta)\exp(j\Phi)\} = \mathrm{Re}\,\overline{\{f(-\delta,t)f^*(0,t)\exp(j\Phi)\}}\,. \tag{7.40}$$

$f_s^*(\delta)$ is the complex conjugate of $f_s(\delta)$, the spatial frequency for the baseline δ. At the output of antenna 3 we have

$$|\underline{f}(x,t)|^2 - \underline{f}_s(0)\,, \tag{7.41}$$

where the second term again corresponds to DC, i.e.

$$f_s(0) = \overline{|f(x,t)|^2} = \overline{\underline{f}(x,t)f^*(x,t)}\,. \tag{7.42}$$

The low frequency fluctuations are now brought to the central laboratory via cables, multiplied and the time average formed. The result is

$$r(x,\delta,\Phi) = \overline{\{\mathrm{Re}(\underline{f}(-\delta,t)\underline{f}^*(0,t)\exp(j\Phi)\} - \mathrm{Re}\,\{f_s^*(\delta)\exp}$$

$$\overline{(j\Phi)\}\{|\underline{f}(x,t)|^2 - f_s(0)\}} =$$

$$= \mathrm{Re}\,\overline{\{\underline{f}(-\delta,t)\underline{f}^*(0,t)\exp(j\Phi)\}\,\underline{f}(x,t)\underline{f}^*(x,t)} - f_s(0)\,\mathrm{Re}\,\overline{\{\underline{f}(-\delta,t)}.$$

$$\overline{.\underline{f}^*(0,t)\exp(j\Phi)\}} - \overline{\mathrm{Re}\,\{f_s^*(\delta)\exp(j\Phi)\underline{f}(x,t)\underline{f}^*(x,t)}} +$$

$$+ f_s(0)\,\mathrm{Re}\,\{f_s^*(\exp(j\Phi)\}\,, \tag{7.43}$$

as the DC component cannot be averaged. By use of the trigonometrical identity

$$\cos \Phi = (1/2)\{\exp(j\Phi) + \exp(-j\Phi)\} \tag{7.44}$$

the real part of (7.43) may be rewritten using (7.40) and (7.42)

$$
\begin{aligned}
r(x,\delta,\Phi) &= \overline{(1/2)\underline{f}(-\delta,t)\underline{f}^*(0,t)\underline{f}(x,t)\underline{f}^*(x,t)\exp(j\Phi)} + \\
&\quad + (1/2)\overline{\underline{f}^*(-\delta,t)\underline{f}(0,t)\underline{f}^*(x,t)\underline{f}(x,t)\exp(j\Phi)} - \\
&\quad - f_s(0)\,\mathrm{Re}\,\{f_s^*(\delta)\exp(j\Phi)\} - \mathrm{Re}\,\{f^*(\delta)\exp(j\Phi)\}\,f_s(0) + \\
&\quad + f_s(0)\,\mathrm{Re}\,\{f_s^*(\delta)\exp(j\Phi)\}\ .
\end{aligned}
\tag{7.45}
$$

With (7.36)

$$
\begin{aligned}
(r,\delta,\Phi) &= (1/2)\overline{\underline{f}(-\delta,t)\underline{f}^*(0,t)\underline{f}(x,t)\underline{f}^*(x,t)}\exp(j\Phi) + \\
&\quad + (1/2)\overline{\underline{f}(-\delta,t)\underline{f}^*(x,t)\underline{f}^*(0,t)\underline{f}(x,t)}\exp(+j\Phi) + \\
&\quad + (1/2)\overline{\underline{f}^*(-\delta,t)\underline{f}(0,t)\underline{f}(x,t)\underline{f}^*(x,t)}\exp(-j\Phi) + \\
&\quad + (1/2)\overline{\underline{f}^*(-\delta,t)\underline{f}(x,t)\underline{f}(0,t)\underline{f}^*(x,t)}\exp(-j\Phi) - \\
&\quad - f_s(0)\,\mathrm{Re}\,\{\underline{f}_s^*(\delta)\exp(j\Phi)\} = \\
&= (1/2)\{\underline{f}_s^*(\delta)\exp(+j\Phi) + \underline{f}_s^*(x+\delta)\underline{f}_s(x)\exp(+j\Phi) + \\
&\quad + \underline{f}_s(\delta)f_s(0)\exp(-j\Phi) + \\
&\quad + \underline{f}_s(x+\delta)\underline{f}_s^*(x)\exp(-j\Phi)\} - f_s(0)\,\mathrm{Re}\,\{\underline{f}^*(\delta)\exp(+j\Phi)\} = \\
&= (1/2)\underline{f}_s^*(x+\delta)\underline{f}_s(x)\exp(+j\Phi) + (1/2)\underline{f}_s(x+\delta)\underline{f}_s^*(x)\exp(-j\Phi)
\end{aligned}
\tag{7.46}
$$

$$r(x,\delta,\Phi) = \mathrm{Re}\,\{\underline{f}_s^*(x)\underline{f}_s(x+\delta)\exp(-j\Phi)\}\ . \tag{7.47}$$

If $\Phi = 0$ is assumed, this gives

$$r(x,\delta,0) = \mathrm{Re}\,\{\underline{f}_s^*(x)\underline{f}_s(x+\delta)\} \tag{7.48}$$

or, for $\Phi = \pi/2$

$$r(x,\delta,\pi/2) = \mathrm{Im}\,\{\underline{f}_s^*(x)\underline{f}_s(x+\delta)\}\ . \tag{7.49}$$

Thus we can obtain real and imaginary parts at the interferometer output by making two independent observations with $\Phi = 0$ and $\Phi = \pi/2$, obtaining

$$r(x,\delta) = \underline{f}_s^*(x)\underline{f}_s(x+\delta) \tag{7.50a}$$

or

$$r(x,\delta) = r(x,\delta,0) + jr(x,\delta,\pi/2)\ . \tag{7.50b}$$

In practice $r(x,\delta)$ is measured for several values of x by varying the baseline length of the interferometer. It is advantageous to choose $x = k\delta$, where $k = 1,2,3...n-1,n$.

For every value of k the output function $r(k\delta,\delta)$, which for simplicity is written $r(k\delta)$, is obtained. Using (7.34) for $x = \delta$ we obtain for the spatial frequency spectrum

$$f_s(\delta) = r(0)/f_s(0) \; . \tag{7.51}$$

The spatial frequency component $f_s(0)$ can be found by forming the average value of the square-law detector output function for the element at the position x.

Using equation (7.50) it may be shown that for $n > 1$ we have

$$f_s(n\delta) = \prod_{k=0}^{Q} r\{(n-1-2k)\delta\} \, \bar{r}_n(0) / \left\{ r^* \big((n-2-2k)\delta \big) \right\} \; , \tag{7.52}$$

where

$$Q = (n/2) - 1, \quad \bar{r}_n(0) = f_s(0) \qquad \text{for } n \text{ even} \tag{7.53a}$$

$$Q = (n-1)/2, \quad \bar{r}_n(0) = r_n(0)/f_s(0) \qquad \text{for } n \text{ odd} \; . \tag{7.53b}$$

If the measured ρ are defined as

$$(x) = \arg\{r(x)\} \tag{7.53c}$$

then applying (7.52) the phase of $f_s(n\delta)$ is

$$a(n\delta) = \sum_{k=0}^{n-1} \rho(k\delta) \; . \tag{7.54}$$

Equation (7.54) implies that the phase of the spatial frequency spectrum at $x = n\delta$ is the sum of measured phases of the complex output functions of the compound intensity interferometer determined by using regular steps from $x = 0$ to $x = (n-1)$.

If the baseline between antennas 1 and 2 reduces to zero this system becomes identical to the standard Hanbury-Brown-Twiss system (14)

$$\lim_{\delta \to 0} r(x, \delta) = \lim_{\delta \to 0} f_s^*(x) f(x, \delta) = f_s(x) f_s^*(x) = |f_s(x)|^2 = r(x) \; . \tag{7.55}$$

The output value becomes zero for $\Phi = \pi/2$. The baseline δ is determined by the *sampling theorem* (63,71), which dictates $\delta = \lambda/2$ after determining $f_s(k\lambda/2)$ for $k = 1, 2, ..., n$ the values are introduced into equation (5.21) in order to obtain a map of the brightness temperature distribution of the source.

7.9.3. Very Long Baseline Interferometer and Aperture Synthesis

7.9.3.1. Introduction (VLBI)

VLBI (Very Long Baseline Interferometry) was born in the mid 1960's out of the wish of astronomers to get better angular resolution than single optical telescopes or single radio reflector antennas can achieve. The diffraction limit of a large optical telescope (6 m in diameter) is 0.025 arcseconds. The angular resolution of a telescope is proportional to the wavelength and varies inversely with the diameter of the antenna (see Chapter 1). In the radio domain of the electromagnetic spectrum the resolution is much worse than in the optical so that radioastronomical antennas have to be of large size. But even at high radio frequencies such as 43 GHz a 100 m antenna has an angular resolution of only 17 arcsecs. To further enhance the angular resolution, a possibility is to use two or more telescopes at large separations. This idea has led in the past from twin interferometers (Michelson interferometer Fig. 1.1) up to aperture synthesis using many telescopes spaced at distances as large as the size of the earth allows.

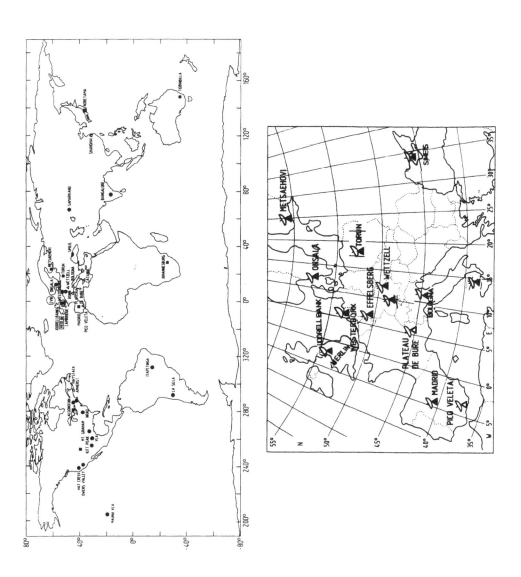

Fig. 7.25.   *VLBI worldwide*

    a.   *Global distribution of radio observatories participating in VLBI experiments;*

    b.   *European radioastronomical stations and some associated stations △.*

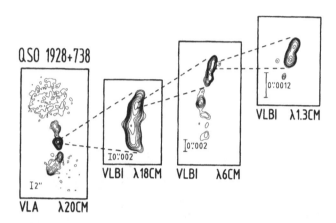

Fig. 7.25.   *VLBI worldwide*
   c.   Quasar 1928+738 at different wavelengths with different an-
        gular resolutions (1983-1986);
   d.   Quasar 3C273 at 2.8 cm (1977-1980).

In VLBI the telescope separations range from a few hundred kilometers up to 9500 kilometers including antenna elements distributed all around the continents. Currently (1988) the record in angular resolution of a ground based observation is held by a VLBI observation at 100 GHz including antennas from Europe, USA and Japan, yielding a resolution of 50 μas (micro arcsecond = 10^{-6} arcseconds) (Baath) (290).

The basic principle of VLBI and connected element interferometry (as discussed earlier in Chapter 6) are very similar, except that the receiving elements now are not connected in "real time". The signals from the radio source are recorded separately at each station on magnetic tape and are later recovered and processed. The recording is performed under the control of independent precise atomic frequency standards (MASERS). The physical independence of these standards is a major difference between VLBI and "connected element" interferometry. The absolute phase information of the recorded VLBI signal is lost normally, but using the method of "closure phase" part of the phase information can be recovered. Further differences compared to connected element interferometry result from the different size, construction and location of each telescope element (the resulting diameter of a twin interferometer is the geometrical mean of the single diameter D_1, D_2) which are normally used as multipurpose radioastronomical instruments not optimized for VLB-interferometry. This necessitates special calibration techniques for VLBI (amplitude and phase calibration and "selfcalibration"). The locations of the participating VLBI-antenna elements on the earth's surface are in some sense a random distribution with baseline lengths ranging from several hundred kilometers to nearly 10^4 km. Figures 7.25 a–d show a global map with positions of single antennas used for cm and mm VLBI, a typical experiment involves a selection of these. Therefore the coverage of the so-called uv- (or projected baseline) plane changes from experiment to experiment and therefore is often quite sparse. The Fourier-transformation to recover the source brightness distribution from the measured interference pattern (the visibility function) consequently requires special imaging procedures (e.g. the "CLEAN" algorithm, Maximum entropy mapping).

7.9.3.2. Applications of VLBI

The basic observables of any interferometer are the fringe phase and fringe amplitude as a function of frequency and time. Derived observables are the fringe rate, which is the time derivative of the fringe phase and the group delay, the frequency derivative of the fringe-phase. These quantities and the interferometer geometry (source positions and telescope coordinates, Table 7.1) represent the information one can obtain. The applications of VLBI to science can be divided mainly into two groups which are not completely independent from each other: the (radio) astronomical and the geophysical application of VLBI.

In radioastronomy continuum VLBI allows us to derive the small-scale brightness distribution of a source, its structure and changes with time. This has led to the discovery of more or less aligned milli-arcsecond scale structures in galaxies and quasars, as well as the discovery of apparent superluminal motion in the jets of some active galactic nuclei. Polarisation VLBI enables astronomers to investigate the source and jet physics using the powerful tool of polarization of electromagnetic radiation.

In contrast to continuum VLBI, spectral line VLBI allows the investigation of the dynamics (position and velocity) and astrophysics of spectral line emitters, such as the various MASERs found in galactic (and extragalactic) molecular clouds and in the atmospheres of stars.

The "phase referencing" method uses a source with known intensity distribution (ideally a point source) to recover the absolute phase information, which is lost in VLBI. This

ID Location	x/m	y/m	z/m	Lon (D)	Lat (D)	D/m
Arecibo	2390500.0000	5564850.0000	1994640.0000	66.3	18.2	305
Crimea	3920644.0	-2563.0	5014000.0	-33.9796	44.3976	22
Darnhall	3829083.7861	16566.7071	5081083.7397	2.5356	53.1567	
Defford	3923438.6111	146912.2011	5009757.6927	2.1444	52.1009	
Dwingeloo	3839344.1181	-430405.1199	5058010.8297			
Effelsberg	4033942.1181	-486993.1199	4900431.8297	-6.8837	50.5252	100
Hatcreek	-2523968.0000	4123511.4000	4147917.5000	121.5	40.8	26
Haystack	1492406.6910	4457267.3300	4296882.1020	71.5	42.6	37
Maryland	1106634.0000	4882910.0000	3938087.1000			
Fort Davis	-1324207.7034	5332028.0100	3232118.1532	103.9	30.6	26
Itapetinga	4033864.4000	4260060.8000	-2495751.6000	1.4	51.1	
Jodrell	3822842.6581	153800.1301	5086287.2197	2.3	53.2	26/76
Madrid	4849015.7000	+360128.3000	4115391.3000	4.2	40.4	64
Medicina	4461363.0000	-919592.5000	4449529.0000	348.4	44.5	32
Metsaehovi	2892579.9681	-1311719.0699	5512640.6897	-24.3932	60.2184	14
Penticton	-2058853.4000	3621457.1000	4814344.7000	119.6	49.3	
Nobeyama	-3871035.2421	-3428098.9311	3724041.2144	221.5	35.9	45
Greenbank	882882.5482	4924484.0469	3944130.8698	79.8	38.4	43
Onsala 20	3370600.4762	-711919.1987	5349831.9259	-11.9264	57.3964	20/26
Onsala 85	3370960.5574	-711467.6611	5349665.1774	348.1	57.4	20/26
Owensvalley	-2409598.1759	4478356.0899	3838602.9331	118.3	37.2	40
Johanb.S.A.	5085424.4181	-2668271.7199	-2768711.9703	332.3	-25.9	
Tidbinbilla	-4460892.6000	-2682358.9000	-3674756.0000	211.0	-35.4	64
Torun	3638607.8000	-1221781.2000	5077149.000	-18.5618	53.0956	15
VLA 27	-1601182.7000	5041978.8000	3554915.7000	107.6	34.1	27 x 25
Westford	1492208.5540	4458131.3290	4296015.8770	71.5	42.6	
Wettzell	4075533.8748	-931737.9380	4801630.4674	-12.8774	49.1452	20
WSRT Array	3828440.6381	-445226.0299	5064923.0797	-6.6334	52.9157	14 x 25

Table 7.1. Location coordinates and data of some VLBI stations.

allows measuring source positions with an absolute positional accuracy of 10^{-3} arcseconds and a relative position accuracy of 10^{-5} arcsecs. In the future it seems therefore possible to establish a position reference system on the sky which is far more accurate than current optical systems (FK4 systems) (136). VLBI also has been used to determine accurately the orbits of earth satellites, to estimate the relative positions of radiotransmitters placed on the moon and planets and to track vehicles on the lunar surface.

The geophysical application of VLBI generally aims at estimating the interferometer geometry from observations of extragalactic sources. One aim of geophysics is to measure the crustal and polar motions of the earth. The measurement of the baseline vectors between two telescopes has led to distance determinations with an accurary of 10. cm and to measurements of crustal plate displacement velocities in the region of 1.-10. cm per year. Further topics of geophysical VLBI-investigations are the measurement of solid earth tides, the changes of the earth rotation axis (precession and nutation), as well as the determination of the earth's polar motion (irregular shifts of the earth's axis) and the observations of the variations of coordinated universal time (UTC), caused by small irregular changes in the angular rotation velocity of our planet.

7.9.3.3. Interferometer Fundamentals

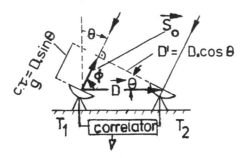

Fig. 7.26.   *Geometry of a two-element interferometer with baseline D including the cross correlator.*

Let us consider the following simplified situation. A source in the far field of two telescopes T_1 at position $\vec{r}_1$ and T_2 at $\vec{r}_2$ emits monochromatic electromagnetic radiation, which may be regarded as an incident plane wave. The telescopes track the source as the earth rotates and they should for simplicity accept the same polarisation, so that we may consider the electric field as a scalar quantity representing only one (linear) polarisation direction. The "baseline" distance of the two telescopes should be $D = |\vec{D}|$, where $\vec{D} = \vec{r}_2 - \vec{r}_1$ is the baseline vector. The angle of the incident radiation is measured with respect to the normal to the vector D and may be called ϑ. Let l be the direction cosine of $\varphi = (90 - \vartheta)$ and $\vec{S}_0$ a vector of unit length pointing from the telescope towards the source. Therefore we have

$$\cos \varphi = \frac{\vec{D} \cdot \vec{S}_0}{|\vec{D}| \cdot |\vec{S}_0|} \tag{7.56}$$

$$l = \sin \vartheta = \cos \varphi . \tag{7.57}$$

The geometrical path difference for the incident plane wavefront with respect to the telescope T_1 and T_2 is $D \cdot \sin \vartheta$ which translates into a geometrical time delay of

$$\tau_g = (D/c)\sin\vartheta \tag{7.58}$$

(with: c = velocity of light) and a geometrical phase shift of

$$\Delta\varphi_g = (2\pi/\lambda_0)\,\vec{S}_0\cdot\vec{D} = (2\pi/\lambda_0)D\cdot 1 \ . \tag{7.59}$$

Whenever the phase difference is an odd (even) multiple of π destructive (constructive) interference appears.

Consider two point sources, one at the direction $l_1 = \sin\vartheta$, the other one at

$$l_2 = \sin\vartheta_2 = \sin(\vartheta + \Delta\vartheta) \ . \tag{7.60}$$

The phase difference between two incoming plane waves from l_1 and l_2 is

$$\delta\varphi = (2\pi\,D/\lambda)\cdot d(\sin\vartheta) = 2\pi\,D_\lambda\,\Delta\vartheta\cos\vartheta \ ; \quad (D_\lambda = D/\lambda_0) \ . \tag{7.61}$$

For a phase difference of $\delta\varphi = 2\pi$ constructive interference appears between the waves from l_1 respectively l_2. This distance between two maxima of an interference fringe pattern determines the angular resolution of an interferometer (see also Fig. 6.2a)

$$\Delta\vartheta = 1/(D_\lambda\cos\vartheta) := 1/u \ , \tag{7.62}$$

where u is usually called the "projected" baseline length as seen from the source. The angular resolution of an interferometer therefore is determined by the length of the baseline, the observing wavelength and the direction to the source. For a source at $\vartheta = 0$ equation (7.62) simplifies to the common expression for the angular resolution A of an interferometer (see also equation (1.1))

$$A \approx \lambda_0/D \ . \tag{7.63}$$

As the earth rotates, ϑ changes with time and the resolution A becomes time dependent. In other words, while ϑ changes from $-90°$ to $90°$ the source is scanned with sinusoidally varying angular resolution. For a particular time t_0 the interferometer is sensitive only to structures in the source of size $\Delta\vartheta \sim 1/u(t_0)$.

Therefore an interferometer may be regarded as a spatial "filter", which successively filters out the different spatial frequencies of the source structure (see chapter 5).

The change of the geometrical phase $\varphi_g := f_F\cdot\tau_g$ with time is called the geometrical or natural fringe rate. From (7.59) follows

$$f_F = (1/2\pi)(\partial\varphi_g/\partial t) = D_\lambda\cos\vartheta(\partial\vartheta/\partial t) = u(\partial\vartheta/\partial t) \ . \tag{7.64}$$

f_F is the frequency shift between the two telescopes due to earth rotation. For an east-west oriented interferometer working at a wavelength of $\lambda_0 = 6$ cm with a baseline length of $D = 8000$ km the natural fringe rate has a maximum value of 9.7 kHz.

The incident radiation induces in the output of each telescope of a multiple telescope interferometer a voltage $V_i(t)$. In crosscorrelation interferometers (see chapter 7.9) these voltages are then multiplied and integrated over a suitable range of time, in other words: the output of a multiplying interferometer is the crosscorrelation function of the antenna signals (see equation (5.13) and Fig. 5.1)

$$r(\tau) = \lim_{T\to\infty}\frac{1}{2T}\int_{-T}^{+T} V_1(t)\,V_2^*(t-\tau)dt :=$$

$$:= \langle V_1(t)\,V_2^*(t-\tau)\rangle \tag{7.65}$$

where the brackets $\langle\ \rangle$ denote the time average, where the asterisk denotes the complex conjugate.

7.9.3.4. Coherence Theory

The understanding of any interference between electromagnetic waves is increased by considering the electrical field at different points in space and time. Let $E(l,t)$ be the (scalar, linearly polarized) electrical field incident from the source. As we observe from the earth in the direction of the source, we consider only the direction cosine 1. We now assume for generality that the radiation of the source is "partially coherent". The frequency spectrum of $E(l,t)$ is given by

$$E(l,t) = \int\limits_{-\infty}^{+\infty} e(l,f)\exp(-j2\pi ft)df \ . \tag{7.66}$$

Therefore for constant l, $e(l,f)$ and $E(l,t)$ form a "Fourier transform pair". In analogy to the frequency spectrum the spatial frequency spectrum can be defined through Fourier transformation (chapter 5)

$$E(l,t) = \int\limits_{-\infty}^{+\infty} e(u,t)\exp(-j2\pi ul)du \ . \tag{7.67}$$

$e(u,t)$ and $E(l,t)$ again are a "Fourier pair". The integration variable u is usually called "spatial frequency" (see chapter 5). The "source coherence function" (mutual coherence function) measures the coherence of the signal in space and time. It is defined as follows

$$\gamma(l_1 l_2 t_1 t_2) := \langle E(l_1,t_1) \cdot E^*(l_2,t_2)\rangle \ , \tag{7.68}$$

which is the ensemble average of the product of the electrical fields over all values of E that satisfy the conditions of the problem. In most practical circumstances a great simplification is made: usually the ensemble averages do not depend on the origin from which time is reckoned. Ensemble averages are then equivalent to time averages and called "stationary". We therefore rewrite the above definition

$$\gamma(l_1 l_2 \tau) = \lim_{T\to\infty} \frac{1}{2T} \int\limits_{-T}^{+T} E(l_1,t)\,E^*(l_2,t-\tau)dt \ =$$

$$= \langle E(l_1,t) \cdot E^*(l_2,t-\tau)\rangle \ , \tag{7.69}$$

where brackets $\langle\ \rangle$ denote time averaging.

As a special case we consider the coherence function at one point, (so $l_1 = l_2$) for $\tau = 0$, which reduces to the intensity (brightness) radiated from the source

$$\gamma(l_1 l_1 0) = \langle |E(l_1)|^2\rangle \sim B(l_1) \ . \tag{7.70}$$

Since $\gamma(l_1 l_2 \tau)$ is not dimensionless, it is often more convenient to deal with the normalized coherence function, the so-called degree of coherence

$$\gamma_N(l_1 l_2 \tau) = \gamma(l_1 l_2 \tau)/\Big(\gamma(l_1,0)\gamma(l_2,0)\Big)^{1/2} \ . \tag{7.71}$$

It is easy to show that γ_N has in its maxima a value of 1.

It is instructive to leave for a moment radiointerferometry and to consider the following simple example of a two "pinhole" interferometer.

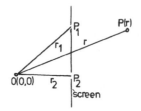

Fig. 7.27. "Pinhole" interferometer.

Let P_1 and P_2 be two pinholes and let P be the point at which the interference phenomena are observed. $\vec{r}_1$, $\vec{r}_2$ and $\vec{r}$ are the position vectors to the corresponding points. Let the field at P_1 be $E_1(\vec{r}_1, t_1)$ and the field that it produces at P be $K_1 E_1(\vec{r}_1, t_1)$, where K_1 is a factor which describes the geometry of the pinhole and all attenuation effects. Similarly, the field at P due to the field at P_2 is $K_2 E_2(\vec{r}_2, t_2)$. The total field in P is then

$$E(\vec{r}, t) = K_1 E_1(\vec{r}_1, t - t_1) + K_2 E_2(\vec{r}_2, t - t_2) \tag{7.72}$$

with

$$t_i = (1/c) |\vec{r}_i| . \tag{7.73}$$

The average intensity I of the total field in P is

$$\langle I(r, t) \rangle = \langle E(\vec{r}, t) E^*(\vec{r}, t) \rangle . \tag{7.74}$$

Inserting equation 7.72 into 7.74 yields after some algebraic operations and the following abbreviations

$$I_1' = |K_1|^2 I_1 = \langle |K_1 E_1(\vec{r}_1, t - t_1)|^2 \rangle \tag{7.75}$$

$$I_2' = |K_2|^2 I_2 = \langle |K_2 E_2(\vec{r}_2, t - t_2)|^2 \rangle \tag{7.76}$$

$$\tau = t_2 - t_1 \tag{7.77}$$

the result

$$I(\vec{r}, t) = I_1' + I_2' + 2\sqrt{I_1' I_2'} \, \mathrm{Re}\left(\gamma_N(\vec{r}_1 \vec{r}_2 \tau) \right) . \tag{7.78}$$

The strength of the interference depends on the magnitude of the coherence function. If the radiation is "completely coherent" the interference should be maximal and $\gamma_N = 1$. For completely incoherent fields $\gamma_N = 0$ and no interference appears, the intensities only add up.

The contrast of the interference fringe pattern is measured by the so-called *visibility*, defined as

$$V = (I_{\max} - I_{\min})/(I_{\max} + I_{\min}) . \tag{7.79}$$

Since

$$I_{\max} = I_1' + I_2'(\pm) 2\sqrt{I_1' I_2'} |\gamma_N| \tag{7.80}$$

it follows that

$$V = 2 \cdot (\sqrt{I_1' I_2'}) |\gamma_N|/(I_1' + I_2') . \tag{7.81}$$

The visibility is therefore proportional to the "degree of coherence". Having understood the principle of coherence an expression may be developed for the crosscorrelation interferometer response. Therefore we first define coherence and incoherence properly (291).

1) The signal from directions l_1 and l_2 are completely coherent (incoherent) if $|\gamma_N(l_1 l_2 \tau)| = 1$ (0) for all values of τ.

2) An extended source is coherent (incoherent) if the radiation from all direction pairs l_1, l_2 within the source are coherent (incoherent).

The "lateral coherence function" $\Gamma(u_1, u_2, \tau)$ is defined as the crosscorrelation of the spatial frequency components of the source

$$\Gamma_{12}(\tau) := \Gamma(u_1, u_2, \tau) = \langle e(u_1, t) e^*(u_2, t - \tau) \rangle , \tag{7.82}$$

where u is the spatial frequency, a coordinate measured in units of wavelengths in a direction normal to $l = 0$. Since $E(l, t)$ and $e(u, t)$ are a Fourier pair, it is easily shown that the source coherence function $\gamma(l_1 l_2 \tau)$ and the lateral coherence function $\Gamma(u_1 u_2 \tau)$ also form a Fourier pair, thus

$$\Gamma(u_1, u_2, \tau) = \int_{-\infty}^{+\infty}\!\!\int \gamma(l_1 l_2 \tau) \exp\left(+j2\pi(l_1 u_1 - l_2 u_2)\right) dl_1 dl_2 . \tag{7.83}$$

This very general expression is simplified in the following way: In radioastronomy the sources are generally assumed to emit incoherent radiation (except for MASERs), which means that for two different points of the brightness distribution of the source $\gamma_N(l_1, l_2, \tau)$ yields only a contribution if $l_1 = l_2$. Thus, using "Diracs Delta function δ" (see equation (7.92), (7.93))

$$\gamma(l_1, l_2, \tau) = \gamma(l_1, \tau) \delta(l_1 - l_2) . \tag{7.84}$$

Inserting this in (7.83) yields, with $u = u_1 - u_2$

$$\Gamma(u, \tau) = \int_{-\infty}^{+\infty} \gamma(l, \tau) \exp(+j2\pi u l) dl \tag{7.85}$$

and in the special case of $\tau = 0$ applying equation (7.70)

$$\Gamma(u, 0) = \int_{-\infty}^{+\infty} \langle |E(l)|^2 \rangle \exp(+j2\pi l u) dl . \tag{7.86}$$

This is the basic relationship between visibility and source brightness, which is known as the *van Cittert-Zernike theorem* (4). The brightness distribution of a source is the Fourier transform of the mutual coherence function. We derived this relation in the special case of one dimension l. For a more general derivation in two dimensions, see: e.g. (292)

$$\Gamma(u, v, 0) = \exp(2\pi j f \delta/c) \int_{-\infty}^{+\infty}\!\!\int B(l, m) \exp\left(-2\pi j(ul + vm)\right) dl dm , \tag{7.87}$$

where $2\pi j\delta/c$ is a phase shift caused by the path difference between the two antennas. For sources in the far field $R \gg D^2/\lambda_0$ (R = distance to the source, D = distance between antennas) of the interferometer this term is small and can be neglected.

7.9.3.5. The Response of a Two-Element Crosscorrelation Interferometer

We discuss now a very general two-element interferometer (see Fig. 7.28).

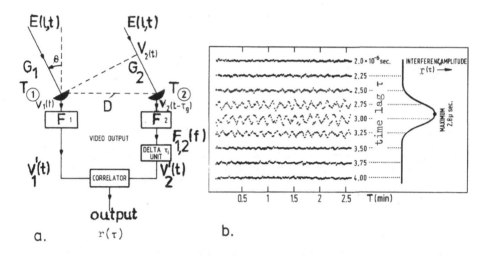

Fig. 7.28.   The response of a twin crosscorrelation
                  interferometer
                  a. Geometry
                  b. Response: fringes $r = f(\tau, T)$, equation (7.97).

The antennas T_1 and T_2 point in direction to the source and track the source. If

$$G_i'(\vartheta, f) := G_i(l, f) \quad \text{with} \quad G_i = 0 \quad \text{for} \quad |l| \geq 1 \tag{7.88}$$

is the angular response pattern (gain) of the i-th antenna, the antenna output voltage becomes

$$V_i(f) = \int\limits_{-\infty}^{+\infty} e(l, f)\, G_i(l, f)dl\,, \quad (i = 1, 2)\,. \tag{7.89}$$

The output of the antenna usually is mixed with a local oscillator and then amplified. The whole receiving system acts as a filter with a frequency response $F_i(f)$. Therefore the receiver output is

$$V_i'(f) = V_i(f)\, F_i(f)\,, \quad (i = 1, 2)\,. \tag{7.90}$$

These signals are now fed into the correlator, which performs crosscorrelation (multiplication with successive time delays and successive integration over time). The correlator output is

$$r(\tau) = \lim_{T \to \infty} \frac{1}{2T} \int\limits_{-T}^{+T} V_1'(t)\, V_2'^*(t' - \tau)\, dt \; . \tag{7.91}$$

Inserting (7.89, 7.90) in (7.91) and using the following property of the delta function (293)

$$\int\limits_{-\infty}^{+\infty} dt \exp\left(2\pi j(f_1 - f_2)\right) t = \delta(f_1 - f_2) \tag{7.92}$$

$$\int\limits_{-\infty}^{+\infty} df\, f(f)\, \delta(f - f') = f(f') \tag{7.93}$$

results in

$$r(\tau) = \int\limits_{-\infty}^{+\infty} df \int\limits_{-\infty}^{+\infty} dl_1 \int\limits_{-\infty}^{+\infty} dl_2 \left[\lim_{T \to \infty} \frac{1}{2T}\, e(l_1 f)\, e^*(l_2 f) \right] \times$$

$$\times\; F_1(f)\, F_2^*(f)\, G_1(l_1 f)\, G_2^*(l_2 f) \exp(-2\pi j f \tau) \; . \tag{7.94}$$

The quantity

$$\gamma(l_1, l_2, f) = \lim_{T \to \infty} \frac{1}{2T} \left[\left(e(l_1 f)\, e^*(l_2 f) \right) \right] \tag{7.95}$$

usually is referred to as the crossspectral function (see section 7.9.3.10, Spectral line VLBI). Its Fourier transform is the source coherence function $\gamma(l_1 l_2 \tau)$ which is also called the mutual coherence function or the crosscorrelation function (ccf) i.e. (see chapter 5, equation (5.13) and 7.9.3.3, equation (7.65))

$$\gamma(l_1, l_2, \tau) = \int\limits_{-\infty}^{+\infty} \gamma(l_1, l_2, f) \exp(-j2\pi f \tau)\, df \; . \tag{7.96}$$

Thus we finally derive

$$r(\tau) = \int\limits_{-\infty}^{+\infty} df \int\limits_{-\infty}^{+\infty} dl_1 \int\limits_{-\infty}^{+\infty} dl_2\, \gamma(l_1, l_2, f)\, F_1(f)\, F_2^*(f)\, G_1(f)\, G_2^*(f) \exp(-2\pi j f \tau) \; . \tag{7.97}$$

This is the general expression for the correlator interferometer output. In the special case of an ideal interferometer (with $F_1 = F_2$, $|F|^2 = 1$, $G_1 = G_2$, $|G|^2 = 1$) looking at a point source equation (7.97) simplifies to equation (7.96). The interferometer measures the crosscorrelation function $r(\tau) = \gamma(l_1 l_2 \tau)$ which is the (frequency) Fourier transform of the crossspectral function.

As mentioned earlier most cosmic radio sources are believed to emit incoherent radiation. Applying therefore equation (7.84) yields in the case of two identical telescopes $G_1 = G_2$, $F_1 = F_2$

$$r(\tau) = \int\limits_{-\infty}^{+\infty} df \int\limits_{-\infty}^{+\infty} dl\, \gamma(l, f)\, |G(l, f)|^2\, |F(l, f)|^2 \exp\left(-2\pi j f \tau(l)\right) \; . \tag{7.98}$$

We now may carry out the frequency integration. If the frequency filter $F(f)$ has a bandpass characteristic with bandwidth Δf centered around a fixed center frequency f_0, and if $\gamma(l, f)$ and $G(l, f)$ do not change significantly within the bandpass we may write

$$r(\tau) = \int\limits_{f-\Delta f}^{f+\Delta f} |F(f')|^2 \exp(-2\pi j f'\tau) df' \int\limits_{-\infty}^{+\infty} \gamma(l, f_0) |G(l, f_0)|^2 \exp(-2\pi j f_0 \tau) dl \ . \tag{7.99}$$

7.9.3.5.1. The fringe washing function The frequency integral in equation (7.99) is a function of the delay between the two telescopes T_1 and T_2 and the shape of the bandpass. This function usually is referred to as the "delay pattern", "bandwidth pattern" or as "fringe washing" function $b(f, \tau)$

$$b(f, \tau) = \int\limits_{\text{Filter}} |F(f)|^2 \exp(-2\pi j f \tau) df \ . \tag{7.100}$$

The fringe washing function describes the change of the interferometer response dependent on the delay τ and the frequency f. In the case of a rectangular bandpass characteristic with width Δf (Fig. 7.29a)

$$|F(f)|^2 = \begin{cases} 1 & \text{if } f_0 - (\Delta f/2) < f < f_0 + (\Delta f/2) \\ 0 & \text{otherwise} \end{cases} \tag{7.101}$$

the fringe washing function becomes

$$b(\Delta f, \tau) = \frac{\sin(\pi \Delta f \tau)}{\pi \Delta f \tau} = \sin c(\pi \Delta f \tau) \ . \tag{7.102}$$

Since $b(\Delta f, \tau) \approx 1$ only if $\pi \Delta f \tau \ll 1$, the interferometer response is significantly different from zero if

$$\tau \ll 1/(\pi \cdot \Delta f) \ . \tag{7.103}$$

This is the temporal coherence condition. Equation (7.103) is a fundamental restriction to interferometry (VLBI). Usually the bandwidth of an interferometer is fixed by the receiving system. A strong interferometer response

$$r(\tau) = b(f', \tau) \int\limits_{-\infty}^{+\infty} \gamma(l, f_0) |G(l, f_0)|^2 \exp(2\pi j f_0 \tau) dl \tag{7.104}$$

is only obtained if the total delay τ between the signals from both interferometer arms is much smaller than the system bandwidth. In a very long baseline interferometer (VLBI) the geometrical delay

$$\tau_g = (D/c) \sin \vartheta \tag{7.105}$$

can be as high as $\sim$30 msec. Due to earth rotation this delay is changing with time. For example we regard the widely used Mark II VLBI interferometer system (294). Its observing bandwidth Δf is 2 MHz (292). To find the fringes (i.e. to satisfy equation (7.103)) in practice the a priori timing should be accurate to at least $100/\Delta f$ which is 50μ sec. This is a factor $\sim$1000. smaller than the geometrical delay. Consequently, it is necessary to introduce an adjustable system delay τ_i in the nonretarded interferometer arm (see Fig. 7.30)

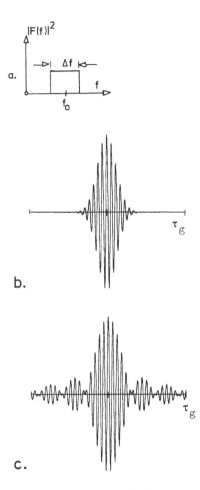

Fig. 7.29. Introduction of a bandpass
a.   The rectangular bandpass (spectral) function
b.   Point source response of an interferometer with a rectangu-
     lar passband (Fig. 7.29)
c.   With Gaussian bandpass.

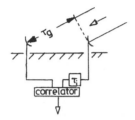

Fig. 7.30. Delays in VLBI (τ_g, τ_i).

$$\tau = \tau_g - \tau_i \ . \tag{7.106}$$

To keep τ constant ($d\tau/dt = 0$) τ_i has to be varied with a rate $d\tau_i/dt = -d\tau_g/dt$. The change of τ_i with time usually is kept under computer control. Condition (7.103) then may be rewritten as

$$\pi\tau\Delta f = \pi(\tau_g - \tau_i)\Delta f = \pi \left| \tau_i - \frac{D}{c} \sin \vartheta \right| \Delta f \ll 1 \ . \tag{7.107}$$

As long as condition (7.107) is satisfied, the fringes are always near the maximum of the envelope of the bandpass. Thee fringes are called "white fringes". The maximum change in the system delay that can be tolerated is from (7.107)

$$\Delta\tau = (D/c) \cos(\vartheta)\Delta\vartheta \ll 1/(\pi\Delta f) \ . \tag{7.108}$$

For the Mark II system ($\Delta f = 2$ MHz) $\Delta\tau \leq 0.16$ μs, for the Mark III system ($\Delta f = 56$ MHz) it is $\Delta\tau \leq 5.7$ ns. The time intervals Δt at which one should compensate follow from (7.108)

$$\Delta t \ll (c/D)(1/\cos \vartheta)(1/\omega_e)(1/\pi\Delta f) \tag{7.109}$$

with the angular velocity $\omega_e = d\vartheta/dt = 2\pi(24 \cdot 60 \cdot 60) = 7.27 \ 10^{-5}$ rad s^{-1} of the earth. For a typical baseline length of $D \approx 8000$ km, a bandwidth of 2 MHz and a value of $\cos \vartheta \approx 1/2\tau_i$ must be changed at least each $\Delta t = 0.16$ seconds.

7.9.3.6. The Interferometer Geometry

Consider now a brightness distribution B in the sky. Let $\vec{S_0}$ be a unit vector pointing to a fixed point P in the celestial sphere having equatorial coordinates α_0, δ_0 respectively h_0, δ_0, where h_0 is the hour angle. The point P can be chosen arbitrarily and for convenience should lie somewhere close to the center of the brightness distribution of the source. Usually P is called the "phase reference point". Let $\vec{S}$ be a similar unit vector pointing to a small element of the brightness distribution near P. This direction is defined by its coordinates $\alpha = \alpha_0 + \Delta\alpha$, $\delta = \delta_0 + \Delta\delta$. Further, we define a right-handed coordinate system (x, y, z) centered at the center of the earth with its z-axis pointing towards the celestial pole ($\delta = 90°$), its x-axis pointing towards $h = 0h$, $\delta = 0°$ and the y-axis pointing towards $h = -6h$, $\delta = 0°$ (Fig. 7.31)

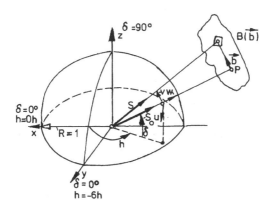

Fig. 7.31. Interferometer geometry in relation to a point P.

We further define a right-handed coordinate system (u, v, w) with the w-axis pointing towards the phase center of the source, therefore being parallel to the vector $\vec{S}_0$. Thus u and v lie in the plane perpendicular to $\vec{S}_0$. The v-axis should point towards the north, the u-axis towards the east. The unit vector $\vec{S}_0$ in direction to the phase center of the source, therefore has the following representation in the x, y, z-system

$$\vec{S}_0 = \begin{pmatrix} S_x^0 \\ S_y^0 \\ S_z^0 \end{pmatrix} = \begin{pmatrix} \cos \delta_0 \cos h_0 \\ -\cos \delta_0 \sin h_0 \\ \sin \delta_0 \end{pmatrix} \ . \tag{7.110}$$

In complete analogy to equation (7.110) the representation of the baseline vector of the interferometer measured in units of wavelength may be written as

$$\vec{D}_\lambda = \vec{D}/\lambda = D_\lambda \begin{pmatrix} \cos D \cos H \\ -\cos D \sin H \\ \sin D \end{pmatrix} = \begin{pmatrix} X_\lambda \\ Y_\lambda \\ Z_\lambda \end{pmatrix} \ . \tag{7.111}$$

The transformation of the elements of a vector represented in the (x, y, z)-system into the (u, v, w)-system is done through applying the following transformation matrix

$$M(\delta_0, h_0) = \begin{pmatrix} \sin h_0 & \cos h_0 & 0 \\ -\sin \delta_0 \cos h_0 & \sin \delta_0 \sin h_0 & \cos \delta_0 \\ \cos \delta_0 \cos h_0 & -\cos \delta_0 \sin h_0 & \sin \delta_0 \end{pmatrix} \ , \tag{7.112}$$

where the elements of the transform matrix are the direction cosines of the (u, v, w)-axis with respect to the (x, y, z)-axis. For example the representation of D in the (u, v, w)-system is given by

$$\vec{D}_\lambda = \begin{pmatrix} u \\ v \\ w \end{pmatrix} = M(\delta_0 h_0) D_\lambda \begin{pmatrix} \cos D \cos H \\ -\cos D \sin H \\ \sin D \end{pmatrix} =$$

$$= D_\lambda \begin{pmatrix} \cos D \sin(h_0 - H) \\ \sin D \cos \delta_0 - \cos D \sin \delta_0 \cos(h_0 - H) \\ \sin D \sin \delta_0 + \cos D \cos \delta_0 \cos(h_0 - H) \end{pmatrix} . \tag{7.113}$$

Similarly, we may represent the vector $\vec{S}$ in the (u, v, w)-system. We define its coordinates as follows

$$\vec{S} = \begin{pmatrix} l \\ m \\ n \end{pmatrix} = M(\delta_0 h_0) \begin{pmatrix} + \cos \delta \cos h \\ - \cos \delta \sin h \\ \sin \delta \end{pmatrix} . \tag{7.114}$$

Finally, we have to find a representation of the vector $\vec{b} = \vec{S} - \vec{S}_0$ in the (u, v, w)-coordinate system. Since $\vec{S}_0$ was defined as a unit vector parallel to the w-axis its representation is just

$$\vec{S}_0 = \begin{pmatrix} l \\ m \\ n \end{pmatrix} = \begin{pmatrix} 0 \\ 0 \\ 1 \end{pmatrix} \Rightarrow \vec{b} = \begin{pmatrix} l \\ m \\ n-1 \end{pmatrix} . \tag{7.115}$$

$\vec{b}$ is a small vector in the plane of the source P, perpendicular to the w-axis, therefore its x-component vanishes. In other words: If l and m are small enough it follows using the following relation of direction cosines

$$l^2 + m^2 + n^2 = 1 \tag{7.116}$$

that $n \approx 1$. This approximation neglects the curvature of the sky over the source extent and is only valid for small sources. Thus $\vec{b}$ has the following representation

$$\vec{b} \approx \begin{pmatrix} l \\ m \\ n \end{pmatrix} \simeq \begin{pmatrix} (h_0 - h) \cos \quad \delta_0 \\ \delta - \delta_0 \\ 0 \end{pmatrix} = \begin{pmatrix} (\alpha - \alpha_0) \cos \quad \delta_0 \\ \delta - \delta_0 \\ 0 \end{pmatrix} . \tag{7.117}$$

Using the vector notation introduced above we now may rewrite the expressions for delay and fringe rate (Equations (7.58, 7.64)). Using (7.110, 7.114) the geometrical delay is given by

$$\tau_g = \frac{1}{c} \vec{D} \cdot \vec{s} = \frac{1}{c} D_\lambda \Big(\sin \delta \sin D + \cos \delta \cos D \cos(h - H) \Big) \tag{7.118}$$

and fringe frequency

$$f_F = \frac{1}{\lambda} \frac{d(\vec{D} \cdot \vec{S})}{dt} = -D_\lambda \cos \delta \cos D \cdot \sin(h - H)\omega_e = -\omega_e \tau_g \tag{7.119}$$

with $\omega_e = \dfrac{dH}{dt}$ the angular velocity of the earth.

7.9.3.6.1. *The interferometer response to an extended source* If we now regard an extended and completely unpolarized and incoherent source and use the coordinate systems introduced in the previous chapter, equation (7.104) becomes in vector notation

$$r'(\tau) = \int |G(\vec{S}, f_0)|^2 \gamma(\vec{S}, f_0) \exp(-2\pi j f_0 \tau) d\vec{S} , \tag{7.120}$$

where we defined for simplicity

$$r(\tau) = b(f, \tau) r'(\tau) \tag{7.121}$$

and $\vec{S}$ as in equation (7.114) (see also Fig. 7.31). Since the geometrical delay is

$$\tau_g = (1/c)\vec{D} \cdot \vec{S} \tag{7.122}$$

and the vector relationship

$$\vec{S} = \vec{S}_0 + \vec{b} \tag{7.123}$$

holds, it follows with $\lambda = c/f_0$, $\vec{b}$ small and nearly perpendicular to $\vec{S}_0$ that

$$f_0\tau = \vec{D}_\lambda \vec{S} = D_\lambda(\vec{S}_0 + \vec{b}) \approx \vec{D}_\lambda \vec{S}_0 + \vec{D}_\lambda \vec{b} \tag{7.124}$$

$$r'(\tau) = \exp(-2\pi j \vec{D}_\lambda \vec{S}_0) \int_{\text{source}} |G(\vec{b}, f_0)|^2 \, \gamma(\vec{b}, f_0) \exp(-2\pi j \vec{D}_\lambda \vec{b}) d\vec{b} \tag{7.125}$$

using equations (7.115, 7.114, 7.113) it follows with $d\vec{b} = dl\,dm/\sqrt{1 - m^2 - l^2}$

$$r'(u, v, w) = \frac{\exp(-2\pi j w)}{\sqrt{1 - l^2 - m^2}} \int |G(l, m, f_0)|^2 \, \gamma(l, m, f_0) \exp\left(-2\pi j(ul + vm)\right) dl\,dm \tag{7.126}$$

where $w = \vec{D}_\lambda \vec{S}_0$ is the geometrical path difference (Fig. 7.32) between the antennas.

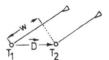

Fig. 7.32. Path length w.

The visibility is defined as

$$V = r'(w = 0) \tag{7.127}$$

and with

$$\gamma(l, m, f_0) = \langle |E(l, m, t)|^2 \rangle \sim B(l, m) \tag{7.128}$$

called the brightness distribution of the source, we obtain finally

$$V(u, f) = \int_{-\infty}^{+\infty} \int |G(l, m)|^2 \, B(l, m) \exp\left(-2\pi j(ul + vm)\right) dl\,dm \ . \tag{7.129a}$$

The visibility is the Fourier transform of the brightness distribution multiplied with the antenna pattern.

A crosscorrelation interferometer measures the visibility function, which is a complex function of the projected baseline vector components (u, v). The brightness distribution of the source can be obtained by applying the inverse Fourier transform of the visibility function.

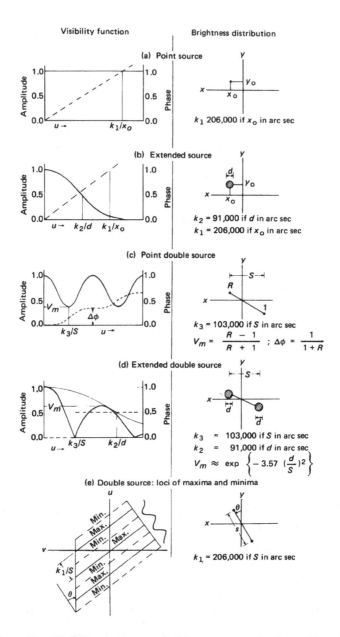

Fig. 7.33.   Visibility function and the appropriate brightness distribution
            models (47).

$$B(l, m).|G(l, m)|^2 = \int_{-\infty}^{+\infty}\!\!\int V(u, v) \exp\Big(+2\pi j(lu + vm)\Big) du\, dv \ . \tag{7.129b}$$

This is the basic principle of imaging in aperture synthesis and interferometry. VLBI uses the method of earth rotation aperture synthesis. As the earth rotates the projected baseline vector as seen from the source

$$\vec{d} = u\vec{e}_u + v\vec{e}_w = \vec{S} \times (\vec{D} \times \vec{S}) \tag{7.130}$$

changes in direction and length. The visibility $V(\vec{d})$ is measured at as many points as possible as a function of the variation of $\vec{d}(t)$, where t denotes for the time dependence introduced by the earth's rotation. The Fourier transform of the complex visibility function $V\Big(u(t), v(t)\Big)$ corresponding to equation (7.129b) finally determines the brightness distribution.

Fig. 7.33 shows examples of the one-dimensional visibility function $V(u)$ and the corresponding brightness distributions.

7.9.3.6.2. The $u - v$-plot The $u - v$-plane is the plane perpendicular to the direction to the phase center of the source. Most VLBI experiments last up to twelve hours. The earth rotates during such an experiment and the projection of the baseline vector to the $u - v$-plane changes with time. Starting from Equation (7.111) and (7.113) we obtain

$$u = D_\lambda \cos D \sin(h_0 - H) \tag{7.131}$$

$$v = D_\lambda\Big(\sin D \cos \delta_0 - \cos D \sin \delta_0 \cos(h_0 - H)\Big) \ . \tag{7.132}$$

Substitution of both equations eliminates the time dependence $(h_0 - H)$ and it follows that

$$(u^2/A^2) + (v - v_0)^2/C^2 = 1 \tag{7.133}$$

where

$$A \ = \ \sqrt{X_\lambda^2 + Y_\lambda^2} \ = \ D_\lambda \cos D \tag{7.134}$$

$$C \ = \ A \sin \delta_0 \quad = \ D_\lambda \cos D \sin \delta_0 \tag{7.135}$$

$$v_0 \ = \ Z_\lambda \cos \delta_0 \quad = \ D_\lambda \sin D \cos \delta_0 \ . \tag{7.136}$$

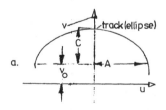

Fig. 7.34a. Elliptical uv-tracks on the celestial sphere.

Equation (7.133) defines an ellipse with semimajor axis A and semiminor axis C centered on the v-axis at $v = v_0$. The length of the arc of the ellipse that is traced out during an observation depends upon the baseline coordinates, the declination of the source and the range of hour angle covered.

The quantity

$$\text{IHA} = h_0 - (H - 6) \quad \text{(calculated in hours)} \tag{7.137}$$

is called interferometer hour angle (IHA). If IHA is zero, u (this is the projection of the baseline in east-west (EW) direction) is maximal, if IHA $= 6\,h$, u is zero. The maximal values of u and v determine the resolution in the east-west and in the north-south direction (NS) respectively

$$A_{EW} \sim 1/u_{max}, \quad A_{NS} \sim 1/v_{max}. \tag{7.138}$$

Consequently, we have

$$u_{max} = D_\lambda \cos D \qquad \text{(at IHA} = 0, \pm 2\,h) \tag{7.139}$$

$$v_{max} = D_\lambda \sin(D + \delta_0) \qquad \text{(at IHA} = \pm 6\,h). \tag{7.140}$$

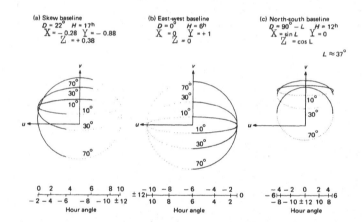

Fig. 7.34b. *The loci of points in the $(u - v)$ plane produced by a tracking interferometer for: a) a skew baseline, b) an east-west baseline, c) a north-south baseline. The loci are drawn for declinations 70°, 30°, and 10°. The solid portion of each curve is given for the hour angle range -6^h to $+6^h$, the dotted portion for the hour angles $+6^h$ to $+18^h$. The hour angle scale, which is a function only of u, is given at the bottom of each diagram. For declinations south of the equator, use the curve with positive declination but flip the curves around $v = v_0$ (46, p. 286).*

The north-south resolution depends on the declination of the source, the east-west resolution does not. For an east-west interferometer ($D = 0$) the east-west resolution is

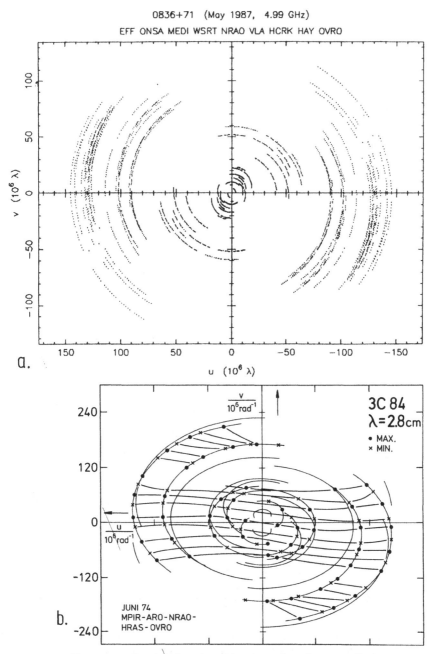

Fig. 7.35.   Tracks of projected baselines (uv-plane in $10^6 \lambda$)
            a. A 9-element interferometer,
            b. A 5-element interferometer.

maximal, but the north-south resolution is minimal and proportional to $\sin \delta_0$. The (uv)-ellipses become circles for a source declination $\delta_0 = 90°$ and they degenerate to straight lines parallel to the u-axis if the source declination becomes $\delta_0 = 0$ (see Fig. 7.34b).

By changing the spacing D_λ of the antennas we obtain several ellipses and thereby fill up the $u - v$ plane. For N different telescope positions one obtains $N(N - 1)/2$ different $u - v$-ellipses. According to equation (7.129) the brightness distribution of a source is reconstructed by sampling the values of the visibility $V(u, v)$ for different values (u, v). The sampling theorem of the Fourier theory (e.g. Bracewell (293)) states that the values of u and v need not be taken any closer than

$$\Delta u, \Delta v < 1/2l_{\max}, 1/2n_{\max} , \tag{7.141}$$

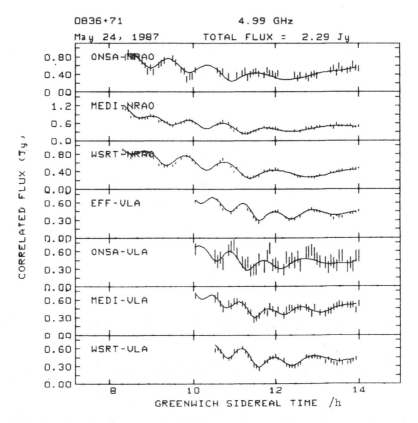

Fig. 7.36.   Correlated flux density of the quasar 0836+71 for some baseli-
            nes of a 9-element VLBI interferometer (Fig. 7.35).

where $l_{\max}$ and $n_{\max}$ are the maximum dimensions of the source. A rough estimate of the number of points that are needed to reconstruct $B(l, m)$ is given by

$$N = (U_{\max}/\Delta U)(V_{\max}/\Delta V) \approx 4D_\lambda^2 \, l_{\max} n_{\max} . \tag{7.142}$$

A property of the Fourier transform of a real function (like $B(l,m)$) is that the visibility obeys the symmetry

$$V(u,v) = V(-u,-v) .\qquad(7.143)$$

Therefore, only half of the N points of the visibility are needed to reconstruct the image. In other words: for a given baseline the visibility is measured on two different $u-v$-tracks which are symmetrical with respect to the origin of the $u-v$-plane. Fig. 7.35 shows the $u-v$-plot of a typical global VLBI experiment with $N=9$ stations. Fig. 7.36 shows as an example the amplitude of the measured visibility function of the source $0836+71$ for different baselines.

7.9.3.7. Single and Double Sideband Systems

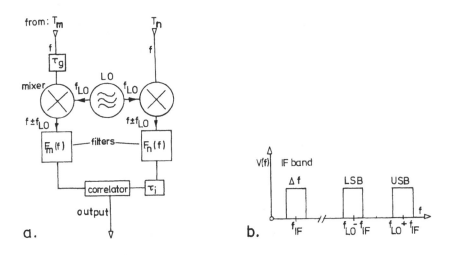

Fig. 7.37.   Application of the heterodyne principle in
radioastronomical receivers
a. block diagram of a receiver ($F_{m,n}=$ filters)
b. SSB and DSB voltage response on the f-scale.

We now develop the ideas introduced in chapter 7.9.3.5 with a more general assumption including the heterodyne technique.

In radioastronomy the frequencies of the signals received at the antennas are changed by mixing with a local oscillator (LO) signal. This technique, referred to as heterodyne frequency conversion, allows the major part of the receiving system of a radiotelescope to work at a fixed intermediate frequency band, which has a number of technical advantages.

Let the antenna output be of frequency f. A local oscillator (LO) may output a signal with frequency f_{LO}. Mixing the local oscillator signal with the antenna signal in a nonlinear device (e.g. a diode) results in output signals composed of the frequencies $(f \pm n \cdot f_{LO})$, where n is an integer. Filtering at $f_{IF} = f - f_{LO}$ removes all higher frequencies ($n > 1$), so that the final signal contains only responses to ($f + f_{LO}$ and $f - f_{LO}$, called the upper (USB) and lower sideband (LSB) respectively, see also chapter 7.9.4 (see Fig. 7.37b). In a single sideband (SSB) system only one of the two possible sidebands is passed by a filter ($F_{m,n}$). In a double sideband system (DSB) both sidebands are used.

7.9.3.7.1. The single sideband system We start by looking at Fig. 7.37a: at telescope m the signal from the radiosources is delayed with respect to telescope n by the geometrical delay $\tau_g = (2\pi/c)\vec{D}\vec{S}$, which is a function of time. To compensate for the time-varying delay τ_g equation (7.58), the instrumental delay τ_i is introduced at telescope n. Usually this time compensation takes place after mixing with the local oscillator signal (frequency f_{LO}, phase ϑ_m respectively ϑ_n) and amplification and filtering. The amplifiers and filters ($F_m(f)$ and $F_n(f)$ are now assumed to be not identical.

The response to an extended source is proportional to the complex visibility function

$$V(u, v) = |V| \exp(i\varphi_v) \tag{7.144}$$

where φ_v is the phase of the visibility. If the visibility function does not vary strongly over the passband, the response of the interferometer at the correlator output may be written (using equation (7.107), (7.125), (7.128))

$$r = \text{Re}\left[\text{const}\,|V| \int_{-\infty}^{+\infty} F_m(f)\, F_n^*(f) \exp\left(i(\varphi_n - \varphi_m - \varphi_v)\right) df\right] \tag{7.145}$$

where φ_n and φ_m describe the phase changes in the signal path of both stations n and m. We now trace the signal through the interferometer branch of station m: at the output of the radio frequency amplifier of the antenna m the signal is delayed with respect to antenna n, giving rise to the phase

$$\varphi_m^{(1)} = +2\pi f(t - \tau_0)\,. \tag{7.146}$$

The mixing of the signal with the local oscillator signal introduces a frequency shift and a local oscillator phase ϑ. Without restriction of generality we regard the lower sideband case: after heterodyne frequency conversion the phase becomes

$$\varphi_m^{(2)} = +2\pi(f - f_{LO})t - 2\pi f\tau_0 + \vartheta_m\,. \tag{7.147}$$

This is seen easily by verifying

$$\cos\omega(t - \tau_0)\cos(\omega_{LO}t = (1/2)\cos[(\omega + \omega_{LO})t - \omega\tau_0] +$$
$$+ (1/2)\cos[(\omega - \omega_{LO})t - \omega\tau_0]\,. \tag{7.148}$$

The first term in the sum represents the USB signal, which is filtered out after the mixing. The second term represents the LSB signal. At station n the phase of the signal after frequency conversion similarly becomes

$$\varphi_n^{(1)} = 2\pi(f - f_{LO})t + \vartheta_n\,. \tag{7.149}$$

The signal now is delayed by an amount of τ_i so that the phase at the correlator input finally becomes

$$\varphi_n^{(2)} = 2\pi(f - f_{LO})(t - \tau_i) + \vartheta_n\,. \tag{7.150}$$

Inserting the phases into equation (7.145) yields with the delay compensation error

$$\Delta\tau = \tau_0 - \tau_i \tag{7.151}$$

$$r = \mathrm{Re}\left[\mathrm{const.}|V| \int_0^\infty F_m(f)\,F_n^*(f)\exp\left(j2\pi\left((f - f_{\mathrm{LO}})\Delta\tau + f_{\mathrm{LO}} \cdot \tau_0\right) + \right.\right.$$

$$\left.\left. + \vartheta_n - \vartheta_m - \varphi_v\right)df\right] \tag{7.152}$$

$$r = \mathrm{Re}\left[\mathrm{const.}|V| \exp\left(j(2\pi f_{\mathrm{LO}}.\tau_0 + \vartheta_n - \vartheta_m - \varphi_v)\right) \int_0^\infty F_m(f)\,F_n^*(f) \cdot\right.$$

$$\left. \cdot \exp(j2\pi(f - f_{\mathrm{LO}})\Delta\tau)df\right] . \tag{7.153}$$

If we define now the complex fringe washing function

$$B_{mn} = |B_{mn}|\exp(i\varphi_B) \tag{7.154}$$

with

$$B(\Delta\tau) = \int_0^\infty F_m(f)\,F_n(f)\exp(j2\pi(f - f_{\mathrm{LO}})\Delta\tau)df \tag{7.155}$$

and the intermediate frequency f_{IF} as the center frequency of the bandpass $F(f)$

$$f_{\mathrm{IF}} = \frac{\displaystyle\int_0^{+\infty} (f - f_{\mathrm{LO}})\,F_m(f - f_{\mathrm{LO}})\,F_n^*(f - f_{\mathrm{LO}})df}{\displaystyle\int_0^{+\infty} F_m(f - f_{\mathrm{LO}})\,F_n^*(f - f_{\mathrm{LO}})df} \tag{7.156}$$

it follows for the lower sideband response of the interferometer

$$r_{\mathrm{LSB}} = |V|\,|B_{mn}(\Delta\tau)| \cos(2\pi f_{\mathrm{LO}}\tau_0 + 2\pi f_{\mathrm{IF}}\Delta\tau + (\vartheta_n - \vartheta_m) - \varphi_v + \varphi_B) . \tag{7.157}$$

Similarly, the response for the USB case is obtained

$$r_{\mathrm{USB}} = |V|\,|B_{mn}(\Delta\tau)| \cos(2\pi f_{\mathrm{LO}}\tau_0 - 2\pi f_{\mathrm{IF}}\Delta\tau + (\vartheta_n - \vartheta_m) - \varphi_v - \varphi_B) . \tag{7.158}$$

$|B_{mn}(\Delta\tau)|$ is the bandpass envelope already introduced in chapter 7.9.3.5.1. Its effect on the response has already been discussed. The quantities $|V|$ and φ_v are the amplitude and phase of the visibility function we wish to measure. The term $2\pi f_{\mathrm{LO}}\tau_0$ is called the "natural fringe phase" and represents the response of the interferometer to a point source measured at frequency f_{LO}. The delay compensation should always maintain $\tau_i \approx \tau_g$. Any delay error $\Delta\tau > 0$ causes the fringe pattern to be phase-shifted by an amount $2\pi f_{\mathrm{IF}}\Delta\tau$. To minimize this phase shift the timing has to be so accurate that

$$\Delta\tau \ll 1/f_{\mathrm{IF}} \text{ holds (for the MK II correlator } f_{\mathrm{IF}} = 0..2 \text{ MHz, } \Delta\tau \ll 0.5 \text{ } \mu s) . \tag{7.159}$$

The interferometer response further is affected by the phases of the local oscillators and their stability. In VLBI, independent local oscillators are used. The phase difference $\vartheta_m - \vartheta_n$ then remains unknown. The phase shift φ_B due to the instrumental properties of the receiving system in principle may be calculated from equation (7.154). However, in a

VLBI system with very different telescopes and receiving systems φ_B essentially remains unknown. In contrast to connected element interferometry, in VLBI the phase of the visibility function cannot be estimated. Special methods therefore are necessary to overcome the phase problem (phase referencing, phase self-calibration, closure phases, see: chapters 7.9.3.10.1.3, 7.9.3.9.2.3, 7.9.3.9.2.1 respectively).

7.9.3.7.2. The double sideband system (DSB) In a double sideband system both sidebands, the lower and the upper sideband are used. From (7.157) and (7.158) the output of a DSB correlation interferometer is

$$r_{\mathrm{DSB}} = r_{\mathrm{LSB}} + r_{\mathrm{USB}} =$$

$$= 2\,|V|\,|B_{mn}|\,(\Delta\tau)\cdot\sin(2\pi f_{\mathrm{IF}}\Delta\tau + \varphi_B)\cdot$$

$$\cdot\cos(2\pi f_{\mathrm{LO}}\tau_0 + (\vartheta_n - \vartheta_m) - \varphi_v)\,. \tag{7.160}$$

The second cosine term corresponds to the SSB response of an ideal interferometer system with perfect delay compensation ($\Delta\tau = 0$) and an ideal phase-stable receiving system ($\varphi_B = 0$). The first sine term describes the nonideal case ($\Delta\tau \neq 0$, $\varphi_B \neq 0$). The fundamental difference to the SSB system is that the fringe frequency term is now independent of $\Delta\tau$ or φ_B (small jumps in $\Delta\tau$ no longer introduce a phaseshift, as is the case in the SSB system). $\Delta\tau$ and φ_B now only affect the amplitude which is

$$|r_{\mathrm{DSB}}| = 2\,|V|\,|B_{mn}(\Delta\tau)|\,\cos(2\pi f_{\mathrm{IF}}\Delta\tau + \varphi_B)\,. \tag{7.161}$$

It is seen that the crosscorrelation fringe amplitude falls more rapidly than it does in the SSB case. Consequently, the required precision in timing is significantly increased. This is the main reason why the MK III-VLBI system needs a much more accurate calculation of delays than the older MK II system which is a single sideband system. Fig. 7.38 shows as an example the interferometer response for a single and a double sideband system as a function of $\Delta\tau$.

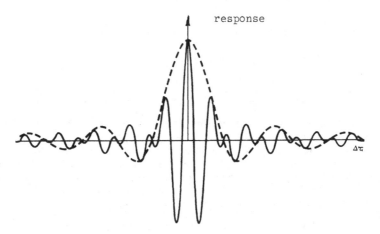

response

$\Delta\tau$

Fig. 7.38.   Fringe amplitude variation as a function of $\Delta\tau$
for a DSB (full line) and SSB (broken line) system
with equal IF responses (292).

7.9.3.8. A Working VLBI System

In the following paragraph we will sketch the signal path through a typical VLBI system: since modern receiver systems are very complex we will restrict ourselves to the most important elements. Our aim is to give the reader an idea how present VLBI systems work and therefore we do not strive for completeness. For more detailed information the reader may refer to (e.g. Meeks 30,31,292).

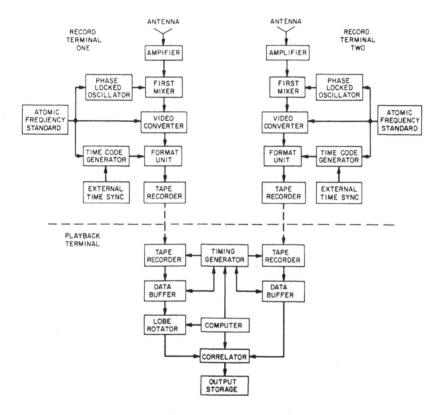

Fig. 7.39a.   Block diagram of a VLBI system (30,31).

A VLBI system with N independent antennas spread over the earth has $N(N-1)/2$ different baselines. Each station records its signal together with accurate time information onto a magnetic tape. The tapes are then brought to a "correlation center", of which one exists in Europe (Bonn, FRG) and several in the USA. Table 7.2 shows a list of these correlation centers. At the correlator the tapes are played back in real time and the data streams of each baseline are correlated independently. To save time all of the correlators can handle more than one baseline at once. Since a typical VLBI experiment has a duration of about twelve hours and may involve ten or more stations the typical time needed for the complete correlation of an experiment may be as large as several hundred hours. To overcome these difficulties correlators which can handle up to 16 stations simultaneously have been built.

Fig. 7.39a represents a general block diagram of a VLBI system. The antennas track the source and the signal is received and amplified using conventional radioastronomical methods. As outlined in chapter 7.9.4, exceptional care must be taken in converting the RF signal to an IF band in order not to destroy the coherence by introducing phase noise into the signal. The local oscillators used in VLBI therefore must be phase locked (PLL) to high-performance frequency standards (rubidium or H MASER clocks). The data are digitized and written to magnetic tapes, together with exact time information. At the correlator the tapes are played back. The data are time-shifted, fringe-rotated and finally crosscorrelated.

	MK II	MK III
Moscow/SU	yes (2)	
Shanghai/China	yes (3)	
NRAO-Charlottesville/Virginia	yes (3)	
JPL/CIT/California	yes (3)	
Max-Planck-Institut für Radioastronomie, Bonn	yes (3)	yes (4)
Haystack/Massachusetts		yes (5)
NRL/Washington, DC		yes (5)
Bologna/Italy	yes (5)	

Table 7.2.   Numbers in parenthesis () give the number of stations (interferometer elements) which can be correlated simultaneously.

7.9.3.8.1. Frequency standards Precise frequency standards of interest for VLBI include crystal oscillators and atomic frequency standards such as rubidium vapour cells, cesium-beam resonators and hydrogen MASERs. Atomic frequency standards incorporate crystal oscillators that are connected to phase-locked loops (PLL) or are frequency-locked to the atomic process using loops with time constants in the range 0.1-1 sec, so that their short term stability is that of the crystal oscillator. The most widely used frequency standard in VLBI is the hydrogen MASER. Modern H-MASERs have a fractional frequency stability up to $f/f_0 \leq 10^{-15}$ for times of 10^{+4} sec (295). At RF frequencies below 1 GHz rubidium standards are sufficient. However, at higher frequencies the use results in a degraded performance of the VLBI system in the reduction of the achievable coherent integration time. The best standards for absolute time measurements are cesium-beam standards, however, their stability in the range 0.1-1000 seconds is not better than that of rubidium standards at a considerably greater cost. Cesium-beam standards are better for time measurement than the rubidium of H-MASER standards because their absolute frequencies are less sensitive to environmental effects ("interference" (296)).

7.9.3.8.2. Time synchronization To find the interference fringes between two stations the time relationship between the signals must be known to within the reciprocal of the signal bandwidth. If the uncertainty is greater than this, the correct time delay can be found by shifting one data stream with respect to the other data stream in time until the fringes are found. This method of trial and error to find the clock offsets between stations is called: "fringe search". It is in any case very time consuming. To minimize the fringe search accurate knowledge of VLBI station clocks is required, for which the following methods are available:

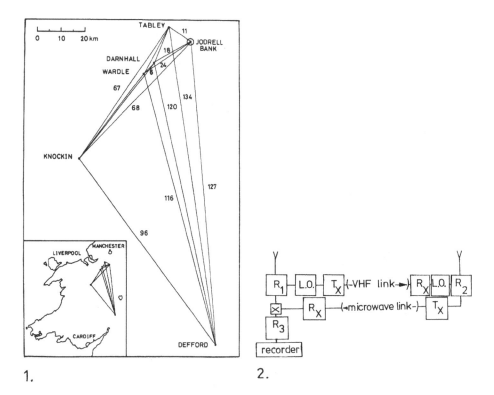

1.   2.

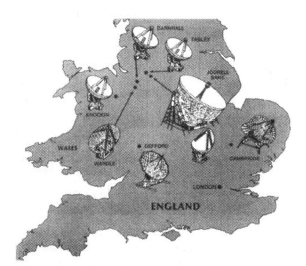

Fig. 7.39b.   Merlin radio antenna array (D_{max} = 134. km, 1980 (298))
1. Geometry; 2. Block diagram; 3. Map.

Fig. 7.39c.   Some VLBI stations/interferometer elements
1. Jodrell Bank Mark II/GB; 2. Nobeyama/Jp; 3. Onsala-
Rao/SW; 4. Pico Veleta/Spain; 5. Simeis-Crimean/SU; 6.
Owens Valley-Caltech/USA.

1) Radio time signals (DCF 77, wwv etc.) enable coarse time synchronisation with an accuracy of about ±1. msec.

2) Transportable cesium-beam clocks sometimes taken to observatories to compare local clocks with national primary standards. The long term stability of these clocks is better than 10^{-13} (1. μsec error in 4. months).

3) Routine clock synchronisation can be obtained from the LORAN C system; "LORAN" is an acronym for Longe Range Navigation, a system originally developed for aircraft and ships. The relative time of arrival of signals from three stations define the observers location on the earth's surface. LORAN C consists of eight master stations, each master station having up to four slave stations. The 100 kHz signal from these stations is available in many parts of the northern hemisphere, allowing clock synchronisation typically to a few μs (292).

7.9.3.8.3. Data recording After the received signals are converted to a videoband (IF-Band) they are filtered and digitized and stored together with the time information on magnetic tape. The digitization takes place in the "formatter" unit. In the MK II VLBI system a two-level quantisation of the signal is used. This kind of digitization is achieved with a "clipper", which is an amplifier working in the saturation regime. Only the sign of the input receiver voltage is conserved. If the voltages at each station are $V_1(t)$ and $V_2(t)$ then the correlation function is

$$r_{12}(\tau) = \langle V_1(t) V_2(t + \tau) \rangle \ . \tag{7.162}$$

The "one-bit-representation" of the voltages after clipping the pulses is

$$x_i = \begin{cases} 1 & V_i > 0 \\ -1 & V_i < 0 \ , \end{cases} \tag{7.163}$$

where $i = 1, 2$. The associated one bit correlation function of the digitized signal is then

$$\rho_{12}(\tau) = \langle x_i(t) x_2(t + \tau) \rangle \ . \tag{7.164}$$

The basic relationship between the autocorrelation function (acf) of a Gaussian signal $V(t)$ and that of the associated clipped signal $x(t)$ is given by the "*van Vleck equation*" (130). The correlation coefficient in units of the flux density S of the observed source is derived from this equation as

$$r_{1,2}(\tau) = S \big((T_{S1}/T_{A1})(T_{S2}/T_{A2}) \big)^{1/2} \sin \big((\pi/2) \rho_{12}(\tau) \big) \tag{7.165}$$

where T_{Si} is the system temperature at station i and T_{Ai} the respective antenna temperature. Equation (7.165) is the basic equation for the amplitude calibration of the correlation coefficient. To calibrate the one-bit crosscorrelation coefficient, the system and antenna temperatures have to be measured at each telescope separately.

It can be shown that two-level quantisation of the signal degrades the signal-to-noise ratio (SNR) compared to the SNR of an analog measurement with equal bandwidth by a factor of $\pi/2$ or 1.7.

After the signal is clipped it is sampled at the Nyquist rate of twice the bandwidth Δf, for example for a bandwidth of 2 MHz the data rate is $4 \cdot 10^6$ Bits/s. This digitization (clipping and sampling) is a method of recording a wideband signal while conserving tape. The recorded data format allows simple clock recovery. To simplify data synchronization, the data are written in blocks (frames) on the tape in addition with a coarse time mark (hours, minutes, seconds) on an auxiliary audio track. The basic fine-time mark then is the frame count which occurs every 1/60 sec (16.67 msec) in the MK II system. Additionally a synchronization pattern is written into the data stream at regular time intervals (every 512 μs in the MK II system).

After formatting (digitization and time encoding) the data are written on a magnetic tape. Presently two different types of tapes are used:

1. The Mark II system uses standard video cassettes (TV). Up to four hours of observation can be recorded on such a tape at each station.

2. In the Mark III system special tapes are used; the recording speed and number of channels can be varied for different experimental setups, for example in "mode A" the system has a bandwidth of 56 MHz, giving 13 minutes of observation on one tape. The length of such a tape is about 3000. m and its weight is several kilograms.

After an experiment is finished, all stations involved send their recorded tapes to the central point of data-reduction—the "VLBI correlation center" (e.g. to the MPIfR in Bonn, F.R.G.), see Table 7.2.

7.9.3.8.4. The correlator center At a VLBI correlation center the data are first recovered and decoded by the playback system. Then the correlator synchronizes the tapes in time and shifts the data streams from each station using a digital buffer. The relative time shift of the data streams takes account of clock offsets and geometrical delays. Since the geometrical delay changes with time due to earth rotation a computer is used to calculate the required delay and natural fringe rate. The frequency of one of the two data streams for a baseline is shifted to compensate clock rate offsets and the earth rotation Doppler shift difference between the two stations. This frequency shift is accomplished by multiplying the data stream by a digital approximation (usually three levels) to the desired sine wave. Actually two shifted data streams are formed, using sine waves that are 90° out of phase. The results of the correlation of both these streams with the data from the other antenna of the baseline form the complex correlation function (ccf).

The complex correlation function as a function of time is then accumulated and written to some archive medium, usually a tape for further processing, either in a general purpose computer (e.g. Convex C1a, 1988) or in a dedicated machine with enhanced signal processing hardware (e.g. a HP computer).

7.9.3.9. Aperture Synthesis Mapping

7.9.3.9.1. Introduction In this chapter we will describe the basic principles which lead from the measured complex visibility function to the brightness distribution of the source.

7.9.3.9.2. Calibration An interferometric array measures the spatial coherence function, called the "complex visibility function" at many discrete locations specified by the projected baseline components (u, v). But before the information about the source contained in the visibility is recovered, the data have to be calibrated. The radio signal from the source passes through all kinds of extraterrestrial media (intergalactic, interstellar and interplanetary medium and dust) and through the earth's atmosphere. After collection by the radio antennas the signal is further modified by the receivers, local oscillators, data digitizers and finally by the correlator. The result is that the observed visibility (i.e. the crosscorrelation function) often shows little resemblance to the true visibility (spatial coherence) function. Amplitude and phase of the measured signal are strongly modified. Therefore amplitude and phase calibration is the process of determining and applying the corrections needed to recover the true visibility function.

Due to the independent local oscillators in VLBI the absolute phase information is lost and the phase is corrupt, whereas connected element interferometers are usually able to measure the absolute phase directly. Phase selfcalibration is therefore of great importance in VLBI mapping. It can, however, also be used to improve map quality from connected-element interferometers. Both systems make use of amplitude selfcalibration (e.g. in the MERLIN network, Fig. 7.39 (298)).

The relationship between the visibility $V_{ij,\text{obs}}$ observed at time t on the baseline $i - j$ and the true visibility $V_{ij,\text{true}}(t)$ can be written very generally as

$$V_{ij,\text{obs}}(t) = G_i(t)\, G_j^*(t)\, G_{ij}(t)\, V_{ij,\text{true}}(t) + a_{ij}(t) + \varepsilon_{ij}(t) \ . \tag{7.166}$$

The terms $G_i(t)$ and $G_j(t)$ represent the effects of the complex gains of the array elements i and j (station gain factors G), the term $G_{ij}(t)$ represents the non-factorable part of the gain (baseline gain factor). $a_{ij}(t)$ represents an offset term and $\varepsilon_{ij}(t)$ is a pure noise term due to thermal receiver noise. Usually, the design of the interferometer and the data processing yields baseline terms G_{ij} and a_{ij} which are small compared to station terms G_i, G_j, so that we may neglect baseline effects and rewrite equation (7.166) as

$$V_{ij,\text{obs}}(t) = G_i(t)\, G_j(t)\, V_{ij,\text{true}}(t) + \varepsilon_{ij}(t) \ . \tag{7.167}$$

In aperture synthesis with phase-connected interferometers (like VLA/USA, Westerbork/NL or MERLIN/GB) calibrator sources near the region to be imaged can be used to solve for the element gains as a function of time. A "calibrator source" must be strong enough to be easily detectable and should have a known structure, preferentially pointlike. In that case $V_{ij,\text{true}}$ is known and from equation (7.167) $G_i(t)$ and $G_j(t)$ can be determined. If the equations are overdetermined, then a least-square technique can be utilized effectively in overcoming the random errors embodied in the factor $\varepsilon_{ij}(t)$.

If the interferometer is redundant, i.e. different pairs of array elements measure the visibility at the same point in the uv-plane, this redundancy can be used in addition to determine $\varepsilon_{ij}(t)$. "Redundant calibration" is currently used at the Westerbork synthesis Radio Telescope (WSRT).

However, in VLBI there is little or no redundancy. Also calibrator sources are rare. With resolutions in the milliarcsecond (m.a.s.) range it is indeed very difficult to find pointlike sources. In addition, the most compact sources are usually variable in flux density and structure. Calibration in VLBI therefore means calibration of the amplitude using conventional measurements of single-antenna system temperatures T_s (see equation (7.165)) and phase self-calibration using "closure phases". If enough stations are involved in a VLB experiment, amplitude self-calibration with closure amplitudes may be of some use. However, any kind of self-calibration requires a good starting model of the sources' brightness distribution (e.g. an ellipse) which is usually not available.

7.9.3.9.2.1. The closure phase Before the introduction of the principle of closure phases in VLBI by Jennison (153) only visibility amplitudes were available. Aperture synthesis imaging, especially making VLBI maps, is only possible with the aid of closure phases: As part of equation (7.167) it follows for the phases

$$\varphi_{ij,\text{obs}}(t) = \varphi_{ij,\text{true}}(t) + \vartheta_i(t) - \vartheta_j(t) + \text{noise} \ , \tag{7.168}$$

where φ_{ij} is the phase of the complex visibility function and ϑ_i, ϑ_j are the station phases. The expression $\vartheta_i - \vartheta_j$ includes all station-dependent phase shifts due to clock errors, local oscillator frequency offsets and frequency noise, delay tracking errors, mismatch in polarization feed angle and propagation effects through the different kinds of media, including the atmosphere (154). If we now sum up the phases $\varphi_{ij,\text{obs}}$ over a triangle of three stations i, j, k, (Fig. 7.40) we obtain the observed "closure phase"

$$
\begin{aligned}
c_{ijk,\text{obs}}(t) &= \varphi_{ij,\text{obs}}(t) + \varphi_{jk,\text{obs}}(t) + \varphi_{ki,\text{obs}}(t) = \\
&= \varphi_{ij,\text{true}}(t) + \varphi_{jk,\text{true}}(t) + \varphi_{ki,\text{true}}(t) + (\text{noise}) = \\
&= c_{ijk,\text{true}}(t) + (\text{noise}) .
\end{aligned}
\tag{7.169}
$$

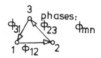

Fig. 7.40. The closure phase loop.

Thus all station-dependent phase errors cancel and the observed closure phase equals the true closure phase (neglecting the noise term), which contains the information about the phase of the true visibility function. The closure phase contains information about the source structure. For any array of N stations with no redundant spacings there are

$$
\big(N(N-1)\big)/2 - (N-1) = (n-1)(n-2)/2
\tag{7.170}
$$

independent closure phases n. Since there are $\big(N(N-1)\big)/2$ visibility phases only $(N-1)$ phases remain unknown when closure phases are used. Since in VLBI the absolute phase information is lost one station baseline phase may be set to zero, so that finally N phases remain unknown. At this point it is necessary to stress, that equation (7.169) only holds under the assumption that the baseline gain factors G are separable in station factors, that means that equation (7.167) is a good approximation of equation (7.166). Whenever baseline-dependent gain factors appear, additional terms in equation (7.169) are added and the phase sum over a triangle no more is closing.

7.9.3.9.2.2. The closure amplitude A closure amplitude can be defined for any loop of four elements of an interferometer array. Again from equation (7.167) (neglecting the noise term) it follows that

$$
\begin{aligned}
A_{ijkl}^{\text{obs}}(t) &= \frac{V_{ij,\text{obs}}(t)\, V_{kl,\text{obs}}(t)}{V_{ik,\text{obs}}(t)\, V_{jl,\text{obs}}(t)} = \\
&= \frac{V_{ij,\text{true}}(t)\, V_{kl,\text{true}}(t)}{V_{ik,\text{true}}(t)\, V_{jl,\text{true}}(t)} = A_{ijkl}(t) .
\end{aligned}
\tag{7.171}
$$

It is seen that the station-dependent gain factors $G_i(t)$ cancel out and the closure amplitude is independent from station gain effects. For N stations there are $\big(N(N-1)\big)/2 - N =$

$N(N-3)/2$ independent closure amplitudes.

7.9.3.9.2.3. Selfcalibration As stated above, closure relations supply only a fraction of
the total number of amplitudes and phases needed to fully constrain the visibility function. The principle in selfcalibration is to use the above relations and a starting model for
the missing data to derive the true amplitude and phase of the visibility. For example,
in phase selfcalibration (also called hybrid mapping) $(n-1)(n-2)/2$ measured closure
phases require $(N-1)$ additional model phases to determine the $N(N-1)/2$ possible
phases of the visibility.

Mathematically a least-squares method is used to minimize the square S of the modulus
of the difference between the observed visibility function $V_{ij,\text{box}}$ and the corresponding
values $V_{ij,\text{model}}$ of a model. The expression that is minimized is

$$S = \sum_{k} \sum_{\substack{ij \\ i \neq j}} \omega_{ij}(t_k) |V_{ij,\text{obs}}(t_k) - G_i(t_k) G_j^*(t_k) V_{ij,\text{mod}}(t_k)|^2 \qquad (7.172)$$

where the $\omega_{ij}(t_k)$ are weighting coefficients usually chosen to be inversely proportional to
the variance of $V_{ij,\text{obs}}$ to give respect to the noise in the data. Boundary conditions for
the solution of equation (7.172) are the closure relationships for amplitude and/or phase,
which are used to give constraints to the model. Once a complete set of calibrated visibility amplitudes and phases is obtained, a first brightness distribution of the source can be
derived, using mapping algorithms like "CLEAN" or "MEM" (see the following chapters
7.9.3.9.3.3, 7.9.3.9.3.4). With that new brightness distribution an improved model is derived, so that iteratively the final source brightness distribution is obtained, see Fig. 7.41.
For VLBI the most important iterative mapping scheme is "Hybrid mapping", which is
essentially a procedure using phase selfcalibration and the "CLEAN" algorithm. Other
approaches are Schwab's (298) and Cornwell's and Wilkinson's (299) use of amplitude
phase selfcalibration to obtain the final brightness distribution of the source.

7.9.3.9.3. The Fourier inversion of the visibility function In this paragraph we will outline
the principles of mapping the source brightness distribution using the Fourier transform
relationship. Starting with equation (7.129b).

$$A(l,m).B(l,m) = \int\limits_{-\infty}^{+\infty} \int\limits_{-\infty}^{+\infty} V(u,v) \exp\left(2\pi j(ul+vm)\right) du\, dv \; . \qquad (7.173)$$

The visibility function $V(u,v)$ is the two-dimensional Fourier transform of the sky bright-
ness distribution multiplied with the main beam (or single antenna) pattern $A(l,m) = |G(l,m)|^2$. For simplicity we define

$$J(l,m) = A(l,m) \cdot B(l,m) \; . \qquad (7.174)$$

Thus the image $J(l,m)$ of the source brightness distribution and the visibility function
form a two-dimensional Fourier transform pair

$$J(l,m) \underline{\text{FT}} V(u,v) \; . \qquad (7.175)$$

The direct reconstruction of the image of the source brightness distribution $J(l,m)$ would
be easy, if $V(u,v)$ is sampled on a regular spaced grid of values u_c and v_c, since most is
known about Fourier inversion techniques if the sampling is periodic. However, in aperture
synthesis (especially in VLBI), the sample points (u_k,v_k) where the visibility $V(u_k,v_k)$ is

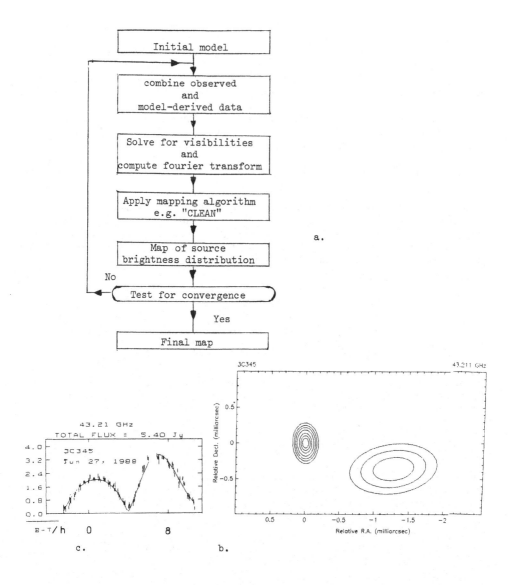

Fig. 7.41. A model of a brightness distribution shown in Fig. 7.41b plotted against a part of the measured visibilities of the source. With sufficient good calibration, convergence against the final model (see scheme indicated above Fig. 7.41a) can be obtained after five to ten iterations.

measured are more or less distributed irregularly over the $u-v$-plane (ellipses) with large areas where no measurements exist. If we define a sampling (or transfer) function $S(u,v)$ which is non-zero only for values (u_k, v_k) where the visibility $V(u_k, v_k)$ is measured we may rewrite equation (7.175).

$$S = \begin{cases} S(u,v) & \text{if } V(u_k,v_k) \text{ exists} \\ 0 & \text{else} \end{cases} \tag{7.176}$$

$$J^D(l,m) = \int\limits_{-\infty}^{+\infty} \int\limits_{-\infty}^{+\infty} S(u,v)\, V(u,v) \exp\Big(2\pi j(ul+vm)\Big)\,du\,dv \ . \tag{7.177}$$

$J^D(l,m)$ is called the "dirty image" or "dirty map". Due to the irregular and incomplete uv-coverage the dirty map is heavily dominated by large sidelobes and radial and irregular structures (since the uv-loci are incomplete ellipses). The dirty map is far away from representing the "true" brightness distribution of the source: Fig. (7.42b) shows an example of a given uv-coverage, Fig. 7.42a the dirty map of a source, for which the deconvolved (cleaned) brightness distribution is shown in Fig. 7.42c.

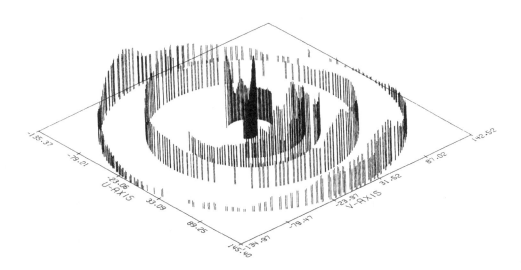

Fig. 7.42. *The clean procedure*
      a.   *$u-v$-coverage in three-dimensional presentation.*
          Extension in z-direction denotes the strength
          of visibility.

If we assume a point source located at the phase reference center of this dirty map, the visibility $V(u,v)$ for such a source is unity in the entire (u,v) plane. The reconstructed image is defined by

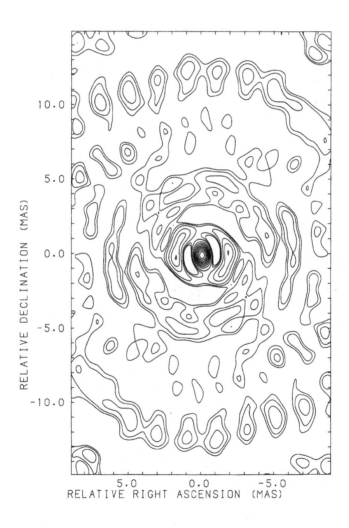

*Fig. 7.42.   The clean procedure
b. dirty beam.*

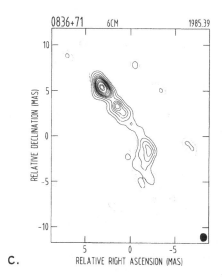

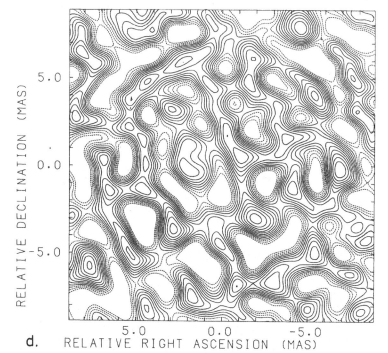

Fig. 7.42.   The clean procedure
c.   final map of the quasar 0836+71
d.   residual map of 0836+71 containing no significant source
     components above the noise level.

$$P(l,m) = \int_{-\infty}^{+\infty}\!\!\!\int S(u,v)\exp\Big(2\pi j(ul+vm)\Big)\,du\,dv\;.\tag{7.178}$$

$P(l,m)$ is the Fourier transform of the sampling function

$$P(l,m)\;\underline{\mathrm{FT}}\;S(u,v)\tag{7.179}$$

and is usually known as the "synthesized beam" of the array or the "point spread function". Since $V(u,v)$ is the Fourier transform of the "true" image $J(l,m)$ (equation (7.175)) and $P(l,m)$ is the Fourier transform of the sampling function, it follows from the convolution theorem that the "dirty" image is the convolution of the beam with the "true" image

$$J^D(l,m) = P(l,m)\otimes J(l,m)\tag{7.180}$$

$$J^D(l,m) = \int_{-\infty}^{+\infty}\!\!\!\int L(l'm')\,P(l-l',m-m')\,dl'dm'\;,\tag{7.181}$$

where $(\otimes)$ denotes the convolution. Equation (7.180) is the familiar result that the observed brightness is the true brightness convolved with the beam of the observing instrument. The above reconstruction of the image usually is called the "principal solution" (see chapter 6 and (15)). It leads to satisfactory results if $P(l,m)$ has a well-defined main beam with small sidelobes. $J^D(l,m)$ is then a smoothed version resembling the "true" image $J(l,m)$. With poorly distributed sample points inside an irregular boundary having large holes, high sidelobes with peculiar structures occur. Not only would the source in the desired field of view smear out into an irregular shape, sources outside the field may also appear inside to confuse the desired image. Further processing must therefore be performed to obtain a more reasonable map, as discussed below. Before we come to the computational methods of the Fourier inversion, at first the sampling and weighting of the visibility function is shortly discussed.

7.9.3.9.3.1. The weighted sampling function Since the sampling function is the Fourier transform of the interferometer beam (in complete analogy to the "transfer function" which is the Fourier transform of the power reception pattern of a single telescope) it controls the beam shape. A more sophisticated version of the sampling function is the "weighted" sampling function, which is used to "fine tune" the interferometer beam shape. From the previous chapter the "ordinary" sampling function $S(u,v)$, which gives only a contribution for measured points (u_k,v_k) in the $u-v$-plane, may be written in terms of the two-dimensional "δ-distribution" ($=$ Dirac's Delta function, see (7.92, 7.93) as

$$S(u,v) = \sum_{k=1}^{M}\delta(u-u_k,v-v_k)\;.\tag{7.182}$$

The weighted sampling function $S'(u,v)$ is a simple extension of this equation according to

$$S'(u,v) = \sum_{k=1}^{M} R_k\,T_k\,D_k\,\delta(u-u_k,v-v_k)\;.\tag{7.183}$$

The weights R_k, T_k and D_k are weights assigned to the visibility points. These data points usually represent time averages of the visibility measurements spaced along the loci of the interferometer $u-v$-tracks.

  R_k is a weight that indicates the reliability of the kth visibility point. Not all interferometer baselines produce similar data quality. The measurements of the visibility for

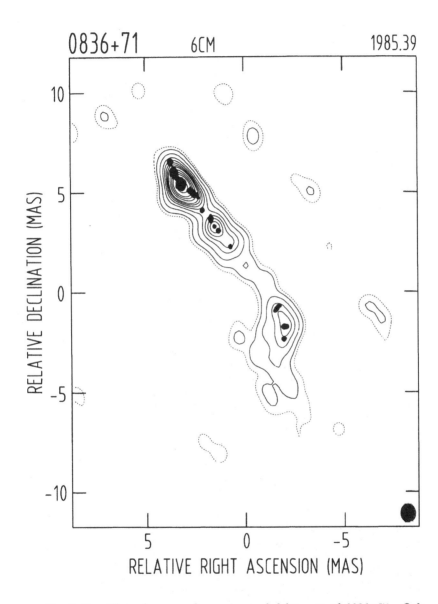

Fig. 7.43.   Cleaned map and superimposed delta map of 0836+71. Only
the strongest delta functions are shown (323). The clean map
is the convolution of the delta functions with a Gaussian-shaped
beam.

different baselines may differ, depending on the telescopes, receiver and local oscillators stabilities and the atmospheric conditions, strongly in that quality. The signal-to-noise ratio of the different visibility points is widely used for the weighting R_k.

The taper T_k and the density weight D_k control the beamwidth and shape. To reduce sidelobes in the image $J(l, m)$ of the brightness distribution of the source, $S'(u, v)$ must be chosen so, as to taper smoothly towards the edges of the measured portion of the $u - v$-plane. For example high weights to large $u - v$-spacings result in a narrow beam with large sidelobes, whereas a taper giving higher weight to smaller spacing produces a somewhat wider beam, but with lower sidelobes. The T_k are used to weight down the data at the outer edge of the $u - v$-converge (T_k usually is a Gaussian function of the radius $r = (u^2 + v^2)^{1/2}$). The density weight D_k takes respect to the strongly varying density of data points in the $u - v$-plane. The concentration of data points near the center of the $u - v$-plane is much higher than in the outer regions. In VLBI D_k usually is taken inversely proportional to the local density of data points within a given cell size $(u + \Delta u, v + \Delta v)$ (uniform weighting).

7.9.3.9.3.2. Gridding This paragraph deals with the numerical calculation of the Fourier transformation of equation (7.175). To reconstruct an image of N^2 map points or pixels (picture elements) roughly N^4 complex multiplications are necessary. For N of order 10^3 this number becomes prohibitive. To increase the computing speed and to decrease the number of arithmetic operations needed, in aperture synthesis imaging the fast Fourier transform (FFT) algorithm is used. The FFT algorithm requires only a few times $N^2 \log_2 N$ operations and not N^4! To apply the FFT method the data have to be regularly (equidistant!) distributed onto a rectangular lattice (usually a power of two number of points along each side) in the $u - v$-plane. Since the observed data seldom lie on such a grid an interpolation procedure must be used to assign visibility values at the grid points. This is the so-called "gridding" of the data. It is the resampling of non-uniformly distributed sample points on an equally spaced grid. In practice gridding is done via convolution, which not only interpolates the data to the desired grid position, but in addition averages (smoothes) data points, which may lie close together.

Let $C(u, v)$ be the desired (for simplicity normalized) convolution function defined on an equally spaced grid with grid points (u_c, v_c). In practice C is taken to be identically zero outside some small bounded region which is typically as small as some (~ 1 to 6) grid cells. The convolution of the weighted visibility function $V'(u, v)$ (equation 7.174, 7.180)

$$V'(u, v) = S'(u, v) V(u, v) =$$

$$= \sum_{k=1}^{M} R_k T_k D_k \delta(u - u_k, v - v_k) V(u_k, v_k) , \qquad (7.184)$$

with $C(u, v)$ then is given through

$$C \otimes V' = \sum_{k=1}^{M} C(u_c - u_k, v_c - v_k) V'(u_k, v_k) . \qquad (7.185)$$

The operation of sampling $C \otimes V'$ at all points onto a regular grid may be represented by the equation

$$V'' = R(C \otimes V') = R\big(C \otimes (S'V)\big) \qquad (7.186)$$

with the resampling function (Bracewell's "sha" function ω) (293)

$$R(u,v) = \omega\big((u/\Delta u),(v/\Delta v)\big) = \sum_{j=-\infty}^{\infty}\sum_{k=-\infty}^{\infty} \delta\big(j - (u/\Delta u), k - (v/\Delta v)\big) \qquad (7.187)$$

where Δu, Δv denotes for the cell size of the grid.

V'' is a linear combination of regularly spaced δ-functions from which the "dirty" image $\tilde{J}^D$ now can be recovered using the Fast Fourier transform algorithm

$$\tilde{J}^D \underline{\text{FFT}} V'' = R\big(C \otimes (S'V)\big) . \qquad (7.188)$$

In analog the beam $\tilde{P}$ is then given through the relationship

$$\tilde{P} \underline{\text{FFT}} R(C \otimes S') . \qquad (7.189)$$

Combining (7.188) and (7.189) with the same arguments used as for the derivation of equation (7.180), we obtain similarly

$$\tilde{J}^D = \tilde{P} \otimes J . \qquad (7.190)$$

The choice of the convolution function $C(u,v)$ determines the strength of the so-called 'aliasing'. Aliasing is the effect that parts of the sky brightness distribution, that lie outside the primary field of view are folded back into the map. To avoid aliasing problems, gridding convolution functions $C(u,v)$ are used whose Fourier transforms drop down very rapidly beyond the edge of the image—e.g. truncated Gaussians or truncated Gaussian tapered sinc functions.

7.9.3.9.3.3. The CLEAN algorithm Starting from equation (7.190) the dirty map is the convolution of the dirty beam with the true brightness distribution

$$\tilde{J}^D(l,m) = \tilde{P}(l,m) \otimes J(l,m) , \qquad (7.191)$$

where the tilt ($\tilde{\ }$) denotes for all corrections (sampling, weighting, gridding) discussed in the previous paragraphs. Knowing $\tilde{J}^D$ and $\tilde{P}$ one can solve for J.

The straightforward procedure for deconvoluting two functions is to take the Fourier transform of the convolution function, which is equal to the product of the Fourier transforms of the two original functions, divide out the Fourier transform of the one known function and finally transform back. However, the "dirty" beam contains large areas of zero, so it is impossible to divide. The "CLEAN" algorithm first introduced by J.A. Hoegbom (300) is the most widely used method to solve equation (7.191). It provides one solution to the convolution equation by representing a radio source by a number of point sources in an otherwise empty field of view. Iteratively the positions and strengths of these point sources are estimated and finally convolved with an artificial "CLEAN"-beam, usually Gaussian beam, see Fig. 7.43, that is, a beam free of sidelobes.

Let $D(l,m)$ be this set of point sources (delta functions)—subtracting these point sources, convolved with the "dirty" beam from the dirty map yields the so-called "residual map", (Fig. 7.42d) $\varepsilon(l,m)$

$$\varepsilon(l,m) = \tilde{J}^D(l,m) - \int_{\text{map}}\!\!\int D(l',m')\,\tilde{P}(l-l',m-m')dl'dm' \qquad (7.192)$$

$$\varepsilon = \tilde{J}^D - \tilde{P} \otimes D , \qquad (7.193)$$

when the residual map becomes zero (exactly: when the features in the residual map represent only noise) the set of point sources convolved with the dirty points completely represent the dirty map. The clean map J_{clean} is derived from the convolution of the sidelobe free clean beam P_{clean} with the set of delta functions

$$J_{\text{clean}} = P_{\text{clean}} \otimes D \tag{7.194}$$

and the final map is defined as the sum of the clean map and the residual map (which represents the residual noise in the map)

$$J_{\text{final}} = J_{\text{clean}} + \varepsilon = P_{\text{clean}} \otimes D + \varepsilon \ . \tag{7.195}$$

To find the correct set of delta functions $D(l,m)$, clean works iteratively. The iteration procedure consists of the following steps:

1) Locate the maximum in the dirty map $\tilde{J}_D$ and determine its amplitude D_i. If desired one may search for peaks only in specified areas of the image, called "clean windows".

2) Convolve the dirty beam $\tilde{J}_D$ with a point source (at this location) of amplitude $\gamma_L D$. ($\gamma_L \leq 1$, γ_L usually is termed the "loop gain").

3) Substract the result of this convolution from the dirty map.

4) Go to 1) unless the residual map $\varepsilon(l,m)$ is below some users specified level.

5) Convolve the accumulated point source model

$$D = \sum_{i=1}^{N} D_i, \quad (N = \text{number of iterations},$$

$$D_i = \text{point source from the } i\text{-th iteration})$$

$$D = \sum_{i=1}^{N} D_i \tag{7.196}$$

with an idealized "clean" beam P_{clean} to obtain a smooth, clean map. The clean beam usually is obtained by fitting a Gaussian to the central lobe of the dirty beam.

6) Add the residual ε to the clean map, to obtain the final map. Fig. 7.43 shows an example of a delta map and the reconstructed clean map. Within the clean mapping procedure the user may interact by setting "clean windows" (see point 1)), by defining the loop gain γ_L corresponding to the complexity of the source, by fixing the number of iterations N and thus determining the cutoff level of the residual map ε. Finally, the "mapper" defines the clean beam (5). This step in principle allows to define beam sizes smaller than the "true" beam size (of the aperture synthesis telescope) given through the maximal $u - v$-spacings. Thus superresolved maps can be made. However, special caution must be taken in interpreting such maps.

Although CLEAN mapping is established since 1974, the theoretical understanding of "CLEAN" is relatively poor. The uniqueness of the solution J_{clean}, especially in the presence of noise, and the noise and error performance of CLEAN still remains to be properly analyzed. Concerning these problems, the interested reader may refer to the special papers of Schwarz 1979 (301).

7.9.3.9.3.4. The maximum entropy method (MEM) The maximum entropy image restauration procedure is an algorithm in which a special function, the entropy of the brightness distribution, is chosen, which, when maximized, fits the data within the noise level and produces a positive image with a compressed range in pixel values. In the literature two forms of such an entropy function have been discussed extensively (see 302). They are

1. $S_1 = - \sum_k (B_k/B_s) \log(B_k/B_s)$ $\qquad$ (7.197)

and

2. $S_2 = - \sum_k \log B_k$, $\qquad$ (7.198)

where $B_k = B_k(l_k, m_k)$ is the brightness distribution and $B_s = \sum_k B_k$. The sums are taken over all discrete values in the map. The expressions for S_1 and S_2 can be derived mathematically using Bayesian statistics or Bose-Einstein-statistics (see 303–305). Maximation of entropy can be regarded as a means of introducing prior information on the sky brightness (such as "positivity" and "smoothness" of the brightness distribution). The entropy expressions are then derived from statistical consideration of the a priori probability distribution on the sky. Maximization of entropy therefore is an attempt to choose the most likely brightness distribution consistent with the measurements. The requirement that each visibility point be fitted exactly is nearly always incompatible with the "positivity" of the MEM map. Consequently, data are usually incorporated with the constraint, that the fit, χ^2, of the predicted visibility to that observed, be close to the expected value

$$\chi^2 = \sum_i \frac{|V(u_i, v_i) - \bar{V}(u_i, v_i)|^2}{\sigma_v^2} \qquad (7.199)$$

where $\bar{V}(u_i, v_i)$ is the predicted visibility and σ_v is the standard deviation of the noise of $V(u_i, v_i)$. The summation is taken overall visibility data.

Mathematically MEM is better understood, but CLEAN is used more widely since it requires less computing time. Both mapping procedures have been shown to work quite well. It seems to be that MEM is a slightly better mapping algorithm for extended and smooth brightness distributions, while CLEAN is better suited for compact structures with a defined smaller number of discrete peaks in the brightness distribution. If possible, both mapping procedures should be used in parallel for data reduction because in general they compliment each other.

7.9.3.10. Further Applications of VLBI

7.9.3.10.1. Spectral line VLBI
7.9.3.10.1.1. Principle In continuum VLBI the cosmic signal is assumed to have a constant amplitude spectrum: the signal is frequency independent within the bandpass of the receivers. Thus the amplitude spectrum of the interferometer output signal is solely determined by the passband characteristics of the receiver and amplifiers (see equation (7.98), chapter 7.9.3.5.1 "Fringe washing function").

For spectral line radio sources the incoming cosmic signal is frequency dependent within the filter passband. For example, this is the case for natural radio MASER emission: the linewidth of MASERs is much smaller than the filter passband.

According to the Wiener-Khinchin relation (see 7.95, 7.96) the crosscorrelation function (7.65)

$r_{12}(\tau, t) = \langle V_1(t) V_2^*(t - \tau) \rangle$ $\qquad$ (7.200)

and the cross-spectral function

$S_{12}(f, t) = V_1(f) V_2^*(f)$ $\qquad$ (7.201)

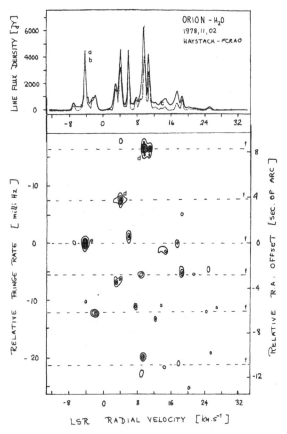

Fig. 7.44.   Fringe rate $\Delta \tau$ frequency distribution of H_2O emission between -17. and
              +25. km/s radial velocity of the 75. km Massachusetts baseline: Hay-
              stack and <u>F</u>ive <u>C</u>ollege <u>R</u>adio-<u>A</u>stronomy <u>O</u>bservation/Amherst at Nov.
              2, 1978.

a.   Autocorrelation-spectrum of Haystack

b.   strong visibility of "small point source": diameter 10^{+14} cm

c.   low "visibility" (strong MASER components, diameter $9 \cdot 10^{+14}$ cm)

d.   cross power

e.   reference line (the ratio of cross- and autocorrelation spectrum is the
      "visibility" if only one component in this velocity channel; if several
      components: their "cross power" must be added

f.   as in (308) these lines mark possible "centers of activity"

g.   contour intervals for "cross power": 6, 15, 30, 57 and 100 percent of
      the strongest lines

h.   fringe-rate-resolution (Haystack-Amherst baseline: 75. km) 0.26 MHz
      corresponds to 0.3 arcsec rather accurate in R.A. as for given $u - v$-
      position of the interferometer only sensitivity in R.A., not in DEC.

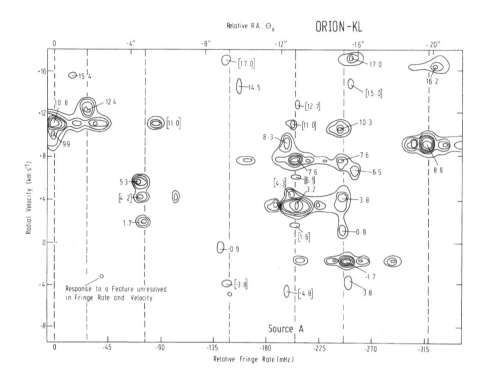

Fig. 7.45. *Fringe rate frequency distribution of H_2O emission of Orion-KL MASER regions between -8. and +17 km/s radial velocity for the Onsala/Sweden-Effelsberg/F.R.Germany baseline in Feb. 2, 1977. The fringe rate $\Delta\tau$ information on the abscissa relative to the reference line at 10.8 km/s is nearly exceptionally in rectascension (RA, see upper horizontal axis). Often lines are of the same radial velocity but of different positions. In these cases, lines, for which a special position could be computed (without parenthesis), for more thin lines with parenthesis the radial velocities are shown. Some of the thick lines indicate even sidelobes, which are an instrumental effect of the Fourier-transformation.*

Further, this map is cleaned from nearly all instrumental effects; contour step units are 1, 5, 10, 15, 25 and 50 percent of the intensity of the strongest line at 10.8 km/s (about 6. kJy).

The vertical dashed lines give hints to centers of activities (240,306). By this figure also rather weak lines can be related to e.g. "source A".

form a Fourier pair (293)

$$r_{12}(\tau, t) = \int_{-\infty}^{-\infty} S_{12}(f, t) e^{j2\pi f \tau} df \qquad (7.202)$$

$$S_{12}(f, t) = \int_{-\infty}^{-\infty} r_{12}(\tau, t) e^{j2\pi f \tau} d\tau . \qquad (7.203)$$

In spectral line VLBI, the spectral analysis is usually achieved with a digital crosscorrelator whose output is the crosscorrelation function $r_{12}(\tau, t)$. This crosscorrelation function is then Fourier transformed to yield the cross-spectral function. The cross-spectral function is a direct measure of the spectrum (the frequency dependence) of the incoming radiation, since the received power $P(f)$ is proportional to the product of the voltages $V_1(f) V_2(f)$.

7.9.3.10.1.2. The phase of the signal We regard a VLBI interferometer with independent local oscillators (frequencies f_{LO}, f_{LO}^2), clock errors τ_1, τ_2 and unknown phase terms $\Theta^1 \Theta^2$: the geometrical delay between the two interferometer arms is

$$\tau_g = \tau_0 + \tau_{RB} + \tau_{atm} , \qquad (7.204)$$

where τ_0 is given in (7.118), τ_{RB} is the delay introduced due to the rotation of the earth during τ_0 (see chapter 7.9.3.10.2.4: Retarded baseline) and τ_{atm} is an atmospheric phase delay. The phases at the input of the correlator are

$$\Phi_1 = 2\pi(f - f_{LO}^1)(t - \tau_1) + \Theta_1 \qquad (7.205)$$

$$\Phi_2 = 2\pi(f - f_{LO}^2)(t - \tau_2) - 2\pi f \tau_g + \Theta_2 . \qquad (7.206)$$

During the correlation the signal from antenna 1 is delayed by τ_0' the estimate of τ_g (delay tracking)

$$\Phi_1 = 2\pi(f - f_{LO}^1)(t - \tau_1 - \tau_0') + \Theta_1 \qquad (7.207)$$

using

$$\Delta f := f_{LO}^1 - f_{LO}^2 \qquad \Delta\Theta = \Theta_1 - \Theta_2$$

$$\Delta\tau_{12} = \tau_1 - \tau_2 \qquad \Delta\tau_g = \tau_0' - \tau_g$$

then

$$\Phi_2 - \Phi_1 = 2\pi\Delta f(t - \tau_1 - \tau_0') - 2\pi f_{LO}^2(\tau_0' + \Delta\tau_{12}) +$$

$$+ 2\pi f(\Delta\tau_{12} + \Delta\tau_g) - \Delta\Theta + 2\pi n . \qquad (7.208)$$

Note, if $\tau_1 = \tau_2$ (no clock errors), $f_{LO}^1 = f_{LO}^2$, identical local oscillators, $\Theta_1 = \Theta_2$ (no phase shift due to LO's) and $\tau_g = \tau_0'$ (perfect delay tracking), then

$$\Phi_{12} = -2\pi f_{LO} \tau_g + 2\pi . n , \qquad n \in I\!N \qquad (7.209)$$

which is independent of the signal frequency, because the delay tracking removes delay-induced phase variations.

In spectral line VLBI two basic methods are of importance

1. phase referencing

2. fringe rate analysis.

The use of a phase reference point allows to measure the delay and the delay errors directly (as long as the reference source and the source are close to each other on the sky ($\Delta\varphi \leq$ some degrees) and the delay introduced from the atmosphere is nearly the same). The delay errors due to the frequency standard and atmosphere are of about one part in 10^{13}, which can cause phase shifts of about $1°$ over a 2 MHz band in 6 hours. This limits the positional accuracy to a few microarcseconds. In practice with the MK II system at 22. GHz, clock and baseline errors dominate and can be measured to an accuracy of about 20.-50. ns, which limits the position accuracy to about 50. microarcseconds (240, 241, 306).

Using only time derivatives $\varphi_{12} = \partial\varphi/\partial t$ (fringe rate), $\dot{\tau}_g = \partial\tau/\partial t$ (delay rate) results in the method of fringe rate analysis. This method is easier to apply than the phase referencing procedure, however, it only provides a measuring accuracy in source positions of order 1. milliarcsecond.

7.9.3.10.1.3. Phase referencing Since local oscillator offsets Δf, phase shifts $\Delta\Theta$ and clock errors $\Delta\tau$ remain largely unknown, it is impossible to recover the absolute phase of the signal from equation (7.208). However, it is possible to get the phase relative to a reference position. We regard e.g. two pointlike MASER sources A and B close to each other on the sky. A and B emit radiation at slightly different frequencies f_A and f_B (different frequencies usually appear because of doppler-shifted radiation, caused by different internal radial motion of the sources). From (7.208) follows for MASER A:

$$\Delta\Phi^a(f_a) = 2\pi\Delta f(t - \tau_1 - \tau_0'^a) - 2\pi f_{LO}^2(\tau_0'^a + \Delta\tau) + 2\pi f_a(\Delta\tau + \tau_0'^a) - \Delta\Theta . \qquad (7.210)$$

Using A as phase reference means correlating both sources with $\tau_g = \tau_g'^a$. Therefore for MASER B

$$\Delta\Phi^b(f_b) = 2\pi\Delta f(t - \tau_1 - \tau_0'^a) - 2\pi f_{LO}^2(\tau_0'^a + \Delta\tau) + 2\pi f_a(\Delta\tau + \tau_0'^a - \tau_g^b) - \Delta\Theta . \qquad (7.211)$$

Taking the phase difference results in

$$\Delta\Phi(b,a) = \Delta\Phi^b - \Delta\Phi^a =$$

$$= 2\pi f_a(\tau_g^a - \tau_g^b) - 2\pi(f_b - f_a)(\tau_0'^a - \tau_g^a + \Delta\tau) . \qquad (7.212)$$

The frequency independent terms $\Delta\Theta$ and $2\pi n$ cancel in equation (7.212). For perfect delay tracking $\tau_g^a \approx \tau_g^{a'}$ and for $\Delta\tau = 0$ the phase difference becomes

$$\Delta\Phi(b,a) = 2\pi f_a(\tau_g^a - \tau_g^b) . \qquad (7.213)$$

If the source is close to its reference feature, then the total delay τ_g may be substituted by its geometrical delay and we obtain with use of equation (7.118) and $\Delta\vec{s} = \vec{s}_a - \vec{s}_b$

$$\Delta\Phi(b,a) = (2\pi f_a/c)\vec{D} \cdot \Delta\vec{S} . \qquad (7.214)$$

7.9.3.10.1.4. The fringe rate method In a nonideal (real) interferometer it is technically difficult to minimize clock errors, that means to make $\tau_0'^a = \tau_g^a$ and $\Delta\tau = 0$. In other words in a real interferometer the second term in equation (7.212) does not vanish totally, thus corrupting the measurable phase difference. In a good approximation the clock errors $\tau_0'^a - \tau_g^a + \Delta\tau$ are time independent for short observation times $\Delta\tau$. Taking therefore the time derivative of the different phase $\Delta\Phi(b,a)$ yields the differential fringe rate

$$\Delta f_{b-a} = \frac{d}{dt}\left(\Delta\varphi(b,a)\right) = 2\pi f_a \frac{d}{dt}(\tau_g^a - \tau_g^b) - 2\pi(f_b - f_a)\frac{d}{dt}(\tau_0'^a - \tau_g) =$$

$$= 2\pi f_a \frac{d}{dt}(\tau_g^a - \tau_g^b) + O(\tau) =$$

$$= 2\pi f_a \frac{d}{dt}(\tau_0^a - \tau_0^b) + O(\tau)\,, \tag{7.215}$$

where we again have made use of the assumption that the atmospheric delays τ_{atm}^a, τ_{atm}^b cancels for sources close to each other on the sky. Under this assumption the differential source vector $\Delta\vec{s}$ may be written (see equation 7.117)

$$\Delta\vec{s} = \begin{pmatrix} (\alpha_b - \alpha_a)\cos\delta_a \\ \delta_b - \delta_a \\ 0 \end{pmatrix} = \begin{pmatrix} \Delta\alpha \\ \Delta\delta \\ 0 \end{pmatrix} \tag{7.216}$$

where (α_a, δ_a) are the coordinates of the reference source A, (α_b, δ_b) the coordinates of the source B.

The representation of the baseline vector in a coordinate system (u, v, w) centered at the reference point A is (equation (7.113))

$$\vec{D} = D_\lambda \begin{pmatrix} \cos D \sin(h_a - H) \\ \sin D \cos\delta_a - \cos D \sin\delta_a \cos(h_a - H) \\ \sin D \sin\delta_a + \cos D \cos\delta_a \cos(h_a - H) \end{pmatrix} \tag{7.217}$$

with (7.214) we obtain for the differential fringe rate

$$\Delta f_{b-a} = (2\pi f_a/c)\frac{d}{dt}\vec{D} \cdot \Delta\vec{s} = -(2\pi\omega_e/c)D_{\lambda a}\cos D\Big(\Delta\alpha\cos(h_a - H) +$$

$$+ \Delta\delta\sin\delta_a\sin(h_a - H)\Big) \tag{7.218}$$

with $\omega_e = dH/dt$, $D_{\lambda a} = D/\lambda_a$.

Since h_a, δ_a, $H(t)$, D are known coordinates of the reference source and the interferometer baseline, the relative position $(\Delta\alpha, \Delta\delta)$ of the MASER source B can be found by fitting equation (7.218) to a series of fringe frequency measurements at various hour angles. This technique was first employed by Moran et al. (30-7) for mapping on OH MASER. Genzel et al. (308) mapped H_2O MASERs at a frequency of 22.GHz. Since the declination of the Orion region is approximately zero, the fringe rate Δf_{b-a} is proportional to $\Delta\alpha$. Fig. 7.44 shows a diagram of the relative fringe rate (positional offset $\Delta\alpha$) versus relative radial velocity (obtained from the frequency offset of the different sources relative to a predefined reference frequency) for water vapour sources in Orion.

7.9.3.10.2. The application of VLBI for Astrometry and Geodesy

7.9.3.10.2.1. *Introduction* In addition to amplitude and phase (which are used for mapping radio sources) group delay $\tau(t)$ and fringe rate $f(t)$ can be measured with a VLB-interferometer. These observables contain the scalar product $\vec{D} \cdot \vec{s}$, which allows to solve for baseline coordinates $(D = x, y, z)$ and source positions (α, δ).

The earth is a spinning body. Its angular momentum vector is not constant with time. Precession and nutation of the earth are measured through the accurate determination of

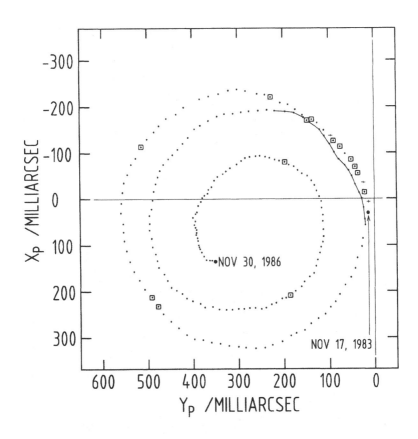

Fig. 7.46.   Polar motion (measured by VLBI (311))
Note that 100 milliarcsec (mas) correspond to 3.2 m.

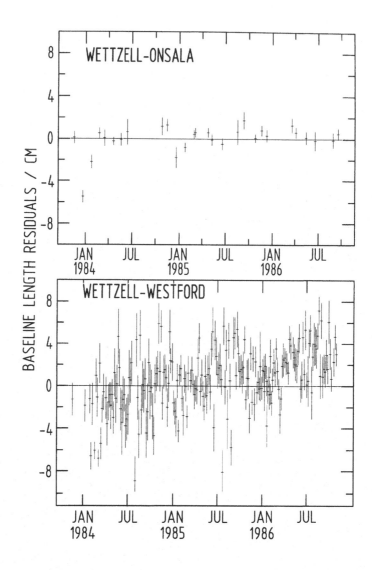

*Fig. 7.47. Increase of baseline length (ΔD/cm) between the VLBI stations
Wettzell/Germany and Onsala/Sweden and Westford/Mass.,
USA in three years (278,289,311).*

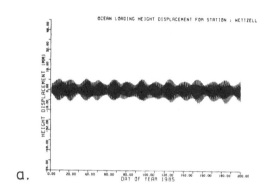

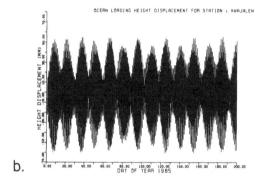

Fig. 7.48.   Vertical (z) displacement due to solid earth tide of the stations:
          a. Wettzell/F.R.Germany
          b. Kwajalen/Finland.

celestial source coordinates. Changes of the rotation axis within the earth's coordinate system are called polar motion. Fig. 7.46 shows the measured polar motion during the last years. Polar motion causes small changes in baseline coordinates (conserving the length of the baseline vector) and therefore is measurable with VLBI. Movements of tectonic earth plates (crustal motion Fig. 7.47) as well as solid earth tides change the baseline vector and its length. As an example Fig. 7.48 shows the vertical displacement of the Wettzell/F.R.G. geodetic VLBI station due to earth tides (310,311).

Changes in the rotation velocity of the earth are caused by polar motion and crustal motion, since both effects change the vector of angular momentum. Fig. 7.49 displays the measured changes of UT1 (UT = universal time) with time.[1]

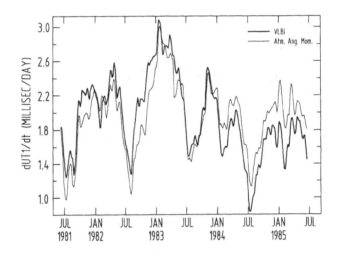

Fig. 7.49.   Changes of UT1
thick line: experimental (VLBI)
thin line: theoretical

The principle of parameter estimation in geodetic VLBI is simple: measure the delay $\tau(t)$ as a function of time (for several sources) and subtract "a priori" model delay terms which correct for:

1) geometrical delay ("a priori" source position and baseline coordinates)

2) radio wave propagation (in the ionosphere and troposphere)

3) instrumental effects (e.g. clock drifts, antenna cable delays, offsets of antenna axis')

4) precession, nutation

5) changes in time (UT1)

6) polar motion

7) earth tides

---

[1]See page 128.

8) delay changes due to source structure

9) relativistic aberration

10) relativistic light deflection

11) other terms.

The a priori model now can be improved iteratively in order to minimize the difference
$$\tau(t) - \tau_M \overset{!}{=} \text{Min}; \text{ where } \tau_M = \sum_{i=1}^{11} \tau_i \text{ is the a priori model delay and } \tau_i \text{ are delay terms}$$
according to the list above. Once $\tau_1 = \tau_0$ (equation (7.117)) is known, the baseline coordinates and the source position can be determined. Repeating this procedure by a long period of time several times per year gives for example baseline coordinates as a function of time which then can be used to determine 6), 7) more accurately and to measure crustal motions.

7.9.3.10.2.2. Sensitivities to delay and fringe rate To get estimates of the sensitivity of delay and fringe rate we shall examine the effects on τ_0 and f_0 (geometrical delay and fringe rate) of small errors in the baseline vector $\vec{D}_\lambda$ equation (7.111), source position vector $\vec{S}$ equation (7.110), polar motion and UT1.[2] This section will show that delay and fringe rate analysis are in large part complementary. After having measured delay τ and fringe rate f_0 we may write

$$\tau = \tau_0 + \tau_M + \Delta\tau , \tag{7.219}$$

where τ_0 is the a priori geometric delay as given by equation (7.118), τ_M is the delay of the a priori model accounting for all other delay terms which are not described by τ_0 (e.g. propagation delay, delay of geodetic motion), and $\Delta\tau$ is the residual delay unaccounted for by the a priori model. A first order expansion of $\Delta\tau$ around the a priori model parameters

$$\Delta\tau = \left(\frac{\partial\tau_0}{\partial X}\Delta X + \frac{\partial\tau_0}{\partial Y}\Delta Y + \frac{\partial\tau_0}{\partial Z}\Delta Z\right) + \left(\frac{\partial\tau_0}{\partial h}\Delta h + \frac{\partial\tau_0}{\partial\delta}\Delta\delta\right) + \Delta\tau_M \tag{7.220}$$

where ΔX, ΔY, ΔZ are the errors in the baseline vector components, Δh (proportional to $\Delta\alpha$), $\Delta\delta$ the errors in source position and $\Delta\tau_M$ the error in modelling the nongeometrical delays.

Similarly the observed fringe rate may be written

$$f = f_0 + \omega\dot{\tau}_M + \Delta f , \tag{7.221}$$

where f is the a priori geometric fringe rate equation (7.119), ω the observing frequency (not a variable, but fixed) and τ_M the assumed delay rate offset (as predicted by the model).

For small errors in the model we may write

$$\Delta f = \left(\frac{\partial f_0}{\partial X}\Delta X + \frac{\partial f_0}{\partial Y}\Delta Y + \frac{\partial f_0}{\partial Z}\Delta Z\right) + \left(\frac{\partial f_0}{\partial h}\Delta h + \frac{\partial f_0}{\partial\delta}\Delta\delta\right) + \omega\Delta\dot{\tau}_M . \tag{7.222}$$

The sensitivity of the geometrical delay to small changes in baseline vector coordinates is described by the partial derivations in equation (7.220)

---

[2] UT1 is a measure of time defined by the earth's rotation angle. An irregular rotation rate of the earth may be measured by determining UT1 as a function of atomic time.

$\partial \tau_0 / \partial X \;=\; (1/c) \cos \delta \cos h$

$\partial \tau_0 / \partial Y \;=\; -(1/c) \cos \delta \sin h$ $\hspace{4cm}$ (7.223)

$\partial \tau_0 / \partial Z \;=\; (1/c) \sin \delta \;.$

These all have an order of magnitude of $1/c$ or about 3 ns/meter, independent of baseline length. Similarly, the fringe rate partials are

$\partial f_0 / \partial X = (-\omega/c) \cos \delta \sin h \,(\partial h/\partial t)$

$\partial f_0 / \partial Y = (-\omega/c) \cos \delta \cos h \,(\partial h/\partial t)$ $\hspace{3cm}$ (7.224)

$\partial f_0 / \partial Z = 0 \;.$

Recalling that $\partial h/\partial t = \omega_e = 7.27 \cdot 10^{-5}$ rad/s and choosing an observing frequency of 8.3 GHz, we find the sensitivity of f_0 to the baseline components X, Y to be about $2 \cdot 10^{-3}$ Hz/m. It is seen that $\partial f_0/\partial Z = 0$. Therefore we note, that fringe rate measurements alone give us no information about the polar baseline component Z. In summary in a VLBI system a timing accuracy of a nanosecond or a fringe rate accuracy of a millihertz is necessary to obtain baseline coordinates with meter accuracy.

To examine the sensitivity of τ_0 and f_0 to small changes in source position we write

$\partial \tau_0 / \partial h = -\partial \tau_0 / \partial \alpha = -\omega/c \,(\partial h)/(\partial t) \cos \delta (X \sin h + Y \cos h)$

$\partial \tau_0 / \partial \delta = -\omega/c \,(\partial h)/(\partial t) \sin \delta (X \cos h - Y \sin h) + Z \cos \delta \;.$ $\hspace{1.5cm}$ (7.225)

The delay sensitivity to source position depends on the baseline length. An order of magnitude calculation for a baseline length of ≈ 8000. km yields a sensitivity of about 130 ns/arc sec = 0.13 ns/mas (1 mas = 10^{-3} arc sec).

For the fringe rate derivatives we obtain

$\partial f_0 / \partial h = -\partial f / \partial \alpha (\omega/c)(\partial h/\partial t) \cos \delta (X \cos h - Y \sin h)$

$\partial f_0 / \partial \delta = -(\omega/c)(\partial h/\partial t) \sin \delta (X \sin h + Y \cos h) \;.$ $\hspace{2cm}$ (7.226)

For 8.3 GHz and a baseline length of ≈ 8000 km the resulting sensitivity is about 80 m Hz/arc sec.

The hour angle h of a source is defined as the difference of the right ascension α of the source and the right ascension of a reference point α_R

$h = \alpha_R - \alpha \;.$ $\hspace{6cm}$ (7.227)

Therefore

$\partial \tau_0 \partial h \;=\; (\partial \tau_0 / \partial \alpha)(\partial \alpha / \partial h) = -(\partial \tau / \partial \alpha) \;=$

$\hspace{1cm} = \; (\partial \tau_0 / \partial \alpha_R)(\partial \alpha_R / \partial \alpha_h) = -(\partial \tau / \partial \alpha_R) \;.$ $\hspace{2.5cm}$ (7.228)

Since α and α_R are directly related they cannot be determined independently. $\partial \tau / \partial \alpha = -(\delta \tau / \partial \alpha_R)$ means, that the origin of right ascension cannot be determined by measurements of τ_0 and f_0.

7.9.3.10.2.3. Calculation of source position (and) or baseline coordinates The geometrical delay (equation 7.118) is

$$\tau_0 = (1/c)\vec{D} \cdot \vec{S} = (1/c)D_\lambda\left(\sin\delta\,\sin D + \cos\delta\,\cos D\,\cos(h - H)\right) . \tag{7.118}$$

The geometrical fringe rate is

$$f_0 = -\omega_e \dot{\tau}_0 = -1.D_\lambda\,\cos\delta\,\cos D\,\sin(h - H)\omega_e . \tag{7.119}$$

The second term of equation (7.118) and the fringe rate (7.119) vary sinusoidally with a period of one sideral day. The remaining term $(1/c)D_\lambda\,\sin\delta\,\sin D$ in (7.118), the so-called "polar component" of the geometrical delay τ_0 is constant with time. In a real VLBI experiment this constant is indistinguishable from a clock synchronisation error, if observations are made only on a single source.

However, if observations are made on several sources (with different declinations δ), equation (7.118) and (7.119) may be used to determine the baseline vector $\vec{D}$. Fringe rate measurements alone only allow to determine the 'equatorial component" $D_\lambda\,\cos D$ of the baseline vector; delay measurements with different souces allow to calculate all three components of $\vec{D}$. Similarly, source coordinates can be obtained for given baseline geometry $(H, D$ known).

A method to determine the unknown parameters from measurements of delay and fringe rate is described in the following.

Inserting equation 7.223 and 7.225 into 7.220 yields

$$\Delta\tau(t) = A_\tau\,\cos h(t) + B_\tau\,\sin h(t) + C_\tau \tag{7.229}$$

with

$$\begin{aligned}
A_\tau &= (1/c)\left((\Delta X - Y\Delta h)\cos\delta - X\Delta\delta\,\sin\delta\right) \\
B_\tau &= -(1/c)\left((\Delta Y + X\Delta h)\cos\delta - Y\Delta\delta\,\sin\delta\right) \\
C_\tau &= (1/c)(\Delta Z\,\sin\delta + Z\Delta\delta\,\cos\delta) + \tau_M .
\end{aligned} \tag{7.230}$$

It is seen that $\Delta\tau(t)$ is a sinoid in $h(t)$ with an offset C_τ. Thus, three parameters (amplitude, phase, offset equivalent to A_τ, B_τ, C_τ) can be measured for any source by observing it for approximately 12. hours. If m sources are observed, $3m$ quantities are obtained. The number of unknown parameters required to specify the m positions, the baseline D, and the residual delay τ_M is $(2m + 3)$ (the right ascension of one source being arbitrarily chosen). Therefore observations on 3 sources are the minimum necessary to determine all the unknown parameters. Usually many more than three sources are observed to get redundant information. A least square fit approach for the solution of equation 7.229 uses this redundancy to increase the accuracy of the parameters obtained.

In a similar fashion equation (7.224) and (7.226) may be used in equation (7.222) to write the fringe rate error as

$$\Delta f(t) = A_f\,\cos h(t) + B_f\,\sin h(t) + C_f \tag{7.231}$$

with

$$\begin{aligned}
A_f &= -(\omega/c)(\partial h/\partial t)\left((\Delta Y + X\Delta h)\cos\delta - Y\Delta\delta\,\sin\delta\right) \\
B_f &= -(\omega/c)(\partial h/\partial t)\left((\Delta X + Y\Delta h)\cos\delta - X\Delta\delta\,\sin\delta\right) \\
C_f &= \omega\dot{\tau}_M .
\end{aligned} \tag{7.232}$$

Comparison with (7.232) shows immediately that

$$A_f = \omega(\partial h/\partial t)B_\tau$$
$$B_f = -\omega(\partial h/\partial t)A_\tau \; . \tag{7.233}$$

These last equations imply that the constants to be solved for in equation 7.229 and 7.231 are equivalent, except that Z and τ_M do not appear in the expression for $\Delta f(t)$. Thus, the addition of fringe rate data to an existing set of delay data does not reduce the minimum number of sources on which observations are necessary in order to solve for all the unknown parameters. Information about source positions and baselines must come from the two parameters A_f, B_f. Therefore unlike the case of delay, where three parameters per source are available, it is not possible to solve for both source and baseline parameters with fringe frequency data. For example, from observations of m sources $(2m + 1)$ quantities are obtained; the total number of unknowns is $(2m+3)$ ($2m$ source parameters, 2 baseline coordinates and C_f). Thus the position of one source must be known to determine the remaining source positions, X, Y and C_f.

Although, measurements of delay and fringe frequency are complementary, they are analysed together in most astrometric experiments. In practice, measurements of delay are generally more accurate because of the noise imposed by the atmosphere. Fringe frequency measurements are sensitive to the time derivative of atmospheric path length (C_f in equation (7.232)) which may change rapidly in a turbulent atmosphere, while the average path length (and thus the average atmosphere delay) is more constant.

7.9.3.10.2.4. The retarded baseline In the previous chapters we have written the geometrical delay in its simplest approximation as $\tau_0 = (1/c)\vec{D} \cdot \vec{S}$. However, the estimate of τ_0 must be accurate enough to ensure that the signal is within the delay ranges of the correlator. As already mentioned in chapter 7.9.3.10.1.2 (equation (7.204)), account must be taken of the fact that the earth moves in the time between the arrival of a wave crest at one station and at another (133,172,309). Therefore the geometrical delay between the two stations is not τ_0 but

$$\tau_0' = \tau_0 + \tau_{RB} = \tau_0(1 + \Delta) \; . \tag{7.234}$$

τ_{RB} is the delay necessary to "retard" the baseline with respect to earth rotation during observation. Let $\vec{r}_1$, $\vec{r}_2$ be the vectors from the center of the earth to each station ($\vec{D} = \vec{r}_2 - \vec{r}_1$). A plane wave reaches the first station at time t_1, the second station at t_2, thus satisfying the "wave equation"

$$2\pi f\left(t_1 - (1/c)\vec{S}\,\vec{r}_1(t_1)\right) = 2\pi f\left(t_2 - (1/c)\vec{S}\,\vec{r}_2(t_2)\right) \; . \tag{7.235}$$

Introducing $\tau_0' = t_2 - t_1$ yields

$$\tau_0' = (1/c)\vec{S}\left(\vec{r}_2(t_1 + \tau_0) - \vec{r}_1(t_1)\right) \tag{7.236}$$

with the Taylor expansion of $\vec{r}_2(t_1 + \tau_0)$

$$\vec{r}_2(t_1 + \tau_0') \simeq \vec{r}_2(t_1) + \frac{d\vec{r}_2(t_0)}{dt} \cdot \tau_0' \; \cdots \cdots \tag{7.237}$$

we may write

$$\tau_0' = (1/c)\vec{S}\,(\vec{D}(t_1) + \dot{\vec{r}}_2(t_1)\tau_0') \; , \tag{7.238}$$

where all quantities are evaluated at t_1. Introducing $\vec{\omega}_e$, the angular velocity vector of the earth $\dot{\vec{r}}_2$ can be written as

$$\dot{\vec{r}}_2 = \vec{\omega}_e \times \vec{r}_2 \ . \tag{7.239}$$

Inserting (7.239) in (7.238) and rearranging gives

$$\tau_0' = (1/c)\,\vec{D}\,\vec{S}\,(1 - \vec{S}(\vec{\omega}_e \times \vec{r}_2)/c)^{-1} \ . \tag{7.240}$$

Taylor expansion of the bracket term yields

$$\tau_0' = (1/c)\,\vec{D}\,\vec{S}\,(1 + \vec{S}\,(\vec{\omega}_e \times \vec{r}_2)/c + ...) \tag{7.241}$$

changing to spherical coordinates according to equation 7.110, 7.111, 7.118 we finally obtain after comparison with (7.234)

$$\tau_0' = \tau_0(1 + \Delta) = \tau_0 + \tau_{\text{RB}} \tag{7.242a}$$

$$\Delta = (1/c)\,\vec{S}\,(\vec{\omega}_e \times \vec{r}_2) =$$

$$= (1/c)\,\omega_e\,r_2\,\cos\varphi_2\,\cos\delta\,\sin(H_2 - h) \tag{7.242b}$$

where h, δ are the source coordinates, φ_2, H_2 are the latitude and hour angle of the baseline vector $\vec{r}_2$, $r_2 = |\vec{r}_2|$ and $\omega_e = |\vec{\omega}_e|$. Δ has a maximum value of $\approx 1.5 \ 10^{-6}$. For a baseline length $D \approx 8000$ km, $\tau_{\text{RB}} = \tau_0 . \Delta = (D/c)\Delta$ may be as high as 40. ns, thus not allowing to ignore this effect.

7.9.3.10.3. Space VLBI

The angular resolution of ground-based VLBI is limited by the size of the Earth (313). The dynamical range and thus the sensitivity of radio source maps, achieved with ground-based VLBI, is mainly determined by the number of uv-points, at which the brightness distribution of the observed source is sampled: the larger the number of observing radiotelescopes, the higher the number of baseline vectors with different lengths and orientations. A logical step in the development of high resolution and high sensitivity VLBI arrays (e.g. VLBA) is the addition of an Earth orbiting radio antenna. This technique was pioneered by Yen et al. (253) in Canada/USA (Fig. 7.50). At an observing frequency of 22 GHz with an apogee height of > 30000 km the resolution of a combined space-to-ground interferometer would be $< 9 \cdot 10^{-5}$ arcsec ($= 90 \ \mu$as).

High resolution mapping of the central engine of the beams and jets of active galactic nuclei (A.G.N.) and quasars could be undertaken. In the core of such radio sources a super massive black hole surrounded by an accretion disk of matter is assumed, which may become directly observable with space VLBI. Furthermore, MASER lines in active star forming regions, accurate positions and position of compact radio sources like quasars, pulsars, late-type stars, etc. can be measured with micro-arcsecond angular resolution.

That space VLBI is technically feasible has been demonstrated recently (317–319) involving a 4.9 m reflector antenna on a TDRSS[3] (Tracking and Data Relay Satellite System) satellite in conjunction with two ground-based 64 m radiotelescopes in Tidbinbilla/Australia and Usuda/Japan at an observing frequency of 2.3 GHz. The longest baseline used in this experiment was 2.15 Earth diameters, giving an angular resolution of better than 1 mas (Fig. 7.52).

In addition to the radio antennas looking at the source in the sky, at least one additional reflector antenna is needed to track the satellite and to transfer a stable reference clock

---

[3]The TDRSS was designed as a communication system employing geosynchronous satellites to maintain contact between low Earth orbit spacecrafts and ground stations.

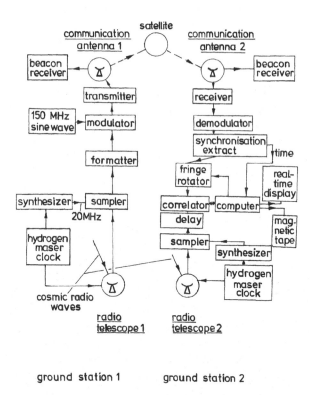

Fig. 7.50. Block diagram of the first satellite VLBI experiment (253).

frequency to the orbiting antenna. This reflector antenna also receives and transfers the source signal from the satellite to a data recording facility. There are plans for future dedicated space VLBI missions: The USA (NASA) plan to launch a 15 m antenna. At the moment (1989) it is uncertain, if this promising project will be carried out in the foreseeable future. The orbit of this satellite named QUASAT (Quasar Satellite) will be excentric with an apogee of 22000–36000 km (time-dependent) and a perigee of 5000 km. Thus the uv-coverage will be excellent. QUASAT operates at the 1.3, 6 and 21 cm wavelength simultaneously, for which special feeds for the reflector elements had to be developed (158–171,314,315,320).

The USSR (INTERKOSMOS) plans to send a 10 m antenna with a very high apogee of 80000 km into space. This satellite (RADIOASTRON) will be equipped with several radio frequency receivers in the range between 1.3 and 92 cm. The primary aim of this mission is to obtain highest resolution at the expense of dynamical range of these images. The Japanese (ISAS) undertake increasing efforts to launch a 10 m reflector antenna with a surface accuracy of $\varepsilon_R \leq 0.5$ mm r.m.s. into orbit (VSOP, VLBI Space Observatory Programme). The orbit, similar as projected for QUASAT, will be extremely excentric (apogee 20000 km, perigee 1000 km) to enable both high sensitivity and high angular resolution aperture synthesis mapping. This satellite will be equipped with three receivers

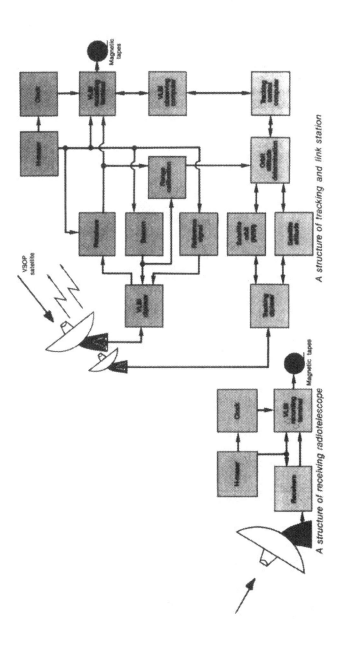

Fig. 7.51.   VSOP satellite: Data link for coherent reception.

for 22 GHz, 5 GHz and 1.7 GHz. A high precision star sensor will enable a pointing accuracy of better than 0.01°. The data transmission rate between orbiter and ground station will be of order ~ 100 Mbits/sec. The launch is planned for 1994/95. The Japanese VSOP is constructed for a lifetime of at least 3 years. A block diagram of the data link for coherent signal reception is shown in Fig. 7.51.

The fact that is of interest for all VLBI and space VLBI projects is that they are a common interdisciplinary and international effort to obtain more knowledge from nature's mystery. Only with space VLBI it is possible to achieve such a high angular resolution, that physical processes hidden in the cores of energetic radio emitters at the edge of our Universe become observable.

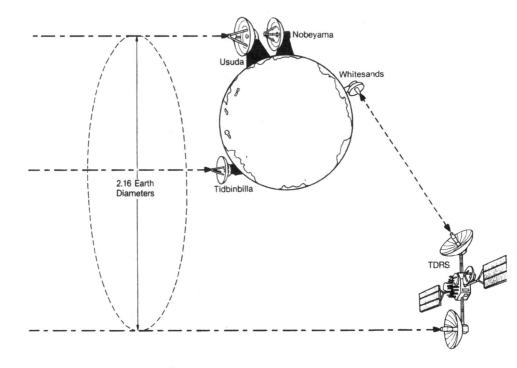

Fig. 7.52. Space VLBI experiment: TDRSS-orbiting (at 2.36 GHz).

7.10. Signal Processing in a Phase Coherent Interferometer

The following outlines the processing of an infinitely narrow band RF signal from a radiosource impinging on a "generalized" two-element interferometer (256,257). This type of interferometer is shown in Fig. 7.53.

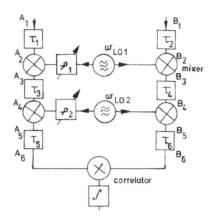

*Fig. 7.53.   Block diagram of the front end of a
mm-wave twin interferometer
(elements: A_1 B_1 (256)).*

It has two frequency conversions and shows delays at all frequency conversion steps and on the two antennas to make it more general. This will be useful, in particular when the case of three antennas is considered. It will be shown in this section that the "general" character is not lost when only two frequency conversions are included and even if this very narrow band case is enlarged later to a finite bandwidth case.

It is assumed that both the lower (L) and upper (U) side bands are received at the radiofrequency stage. It will also be shown later that no generality is lost if a second conversion is assumed either at the lower (L) side band or the upper (U) side band.

ω_{LO1} and ω_{LO2} are the frequencies for the local oscillators. Φ_1 and Φ_2 are two phases added in two phase rotators. These added phases are time-dependent $\Phi_{1,2}(t)$ in order to set the fringe frequency from the natural rate to any desired rate. These phases may also be switched between 0 and π in order to eliminate various offsets. τ_1 and τ_2 are geometrical delays from any arbitrary wave plane of the incoming radiation. τ_3, τ_4, τ_5 and τ_6 are delays within the two receivers; some of them will have to be time variable as shown below.

7.10.1. Signal Processing at a Single Frequency

The signals of the upper and lower sideband are treated separately in the following calculations. Because they are uncorrelated with each other, their correlated products can be simply added to each other as discussed in the section on double sideband processing (7.10.3.1).

We now follow the signals from the impinging plane wave down to the output of the correlator.

At $A1$

$A_U \exp(j\omega_U t)$ in upper sideband   (7.243a)

$A_L \exp(j\omega_L t)$ in lower sideband   (7.243b)

At $B1$

$$B_U \exp(j\omega_U t) \tag{7.244}$$

$$B_U \exp(j\omega_L t) \tag{7.245}$$

At $A2$ (after delay τ_1)

$$A_U \exp\left(j\omega_U(t - \tau_1)\right) \tag{7.246}$$

$$A_L \exp\left(j\omega_L(t - \tau_1)\right) \tag{7.247}$$

At $B3$ (after multiplication by)

$\cos(\omega_{LO1}t - \Phi_1)$ and filtering at $\omega_{IF1} = \omega_U - \omega_{LO1} = \omega_{LO1} - \omega_L$

At $B2$ (after delay τ_2)

$$B_U \exp\left(j\omega_U(t - \tau_2)\right) \tag{7.248}$$

$$B_L \exp\left(j\omega_L(t - \tau_2)\right) \tag{7.249}$$

At $B3$ (after multiplication by)

$\cos(\omega_{LO1}t)$ and filtering at ω_{IF1} $\tag{7.250}$

$$A_U \exp\left\{j(\omega_U - \omega_{LO1})t - \omega_U\tau_1 + \Phi_1\right\} \tag{7.251}$$

$$A_L \exp\left\{j(\omega_{LO1} - \omega_U)t - \omega_L\tau_1 - \Phi_1\right\} \tag{7.252}$$

$$B_U \exp\left\{j(\omega_U - \omega_{LO1})t - \omega_U\tau_2\right\} \tag{7.253}$$

$$B_L \exp\left\{j(\omega_{LO1} - \omega_L)t + \omega_L\tau_2\right\} \tag{7.254}$$

At $A4$: after the delay τ_3

$$A_U \exp\left\{j(\omega_U - \omega_{LO1})(t - \tau_3) - \omega_U\tau_1 + \Phi_1\right\} \tag{7.255}$$

$$A_L \exp\left\{j(\omega_{LO1} - \omega_L)(t - \tau_3) + \omega_L\tau_1 - \Phi_1\right\} \tag{7.256}$$

At $B4$: after the delay τ_4

$$B_U \left\{j(\omega_U - \omega_{LO1})(t - \tau_4) - \omega_U\tau_2\right\} \tag{7.257}$$

$$B_L \left\{j(\omega_{LO1} - \omega_L)(t - \tau_4) + \omega_L\tau_2\right\} \tag{7.258}$$

At $A5$: after the multiplication by $\cos(\omega_{LO2}t - \Phi_2)$ and filtering at the intermediate frequency IF2:

$$\omega_{IF2} = \omega_{IF1} - \omega_{LO2} \tag{7.259}$$

$$A_{\substack{U \\ L}} \exp\left\{j\left(\underbrace{(\pm\omega_L \mp \omega_{LO1} - \omega_{LO2})}_{= \omega\,IF2}t \mp (\omega_{\substack{U \\ L}} - \omega_{LO1})\tau_3 - \omega_{\substack{U \\ L}}\tau_1 \pm \Phi_1 + \Phi_2\right)\right\}. \tag{7.260}$$

At $B5$: after the multiplication by $\cos(\omega_{LO2}t)$ and filtering at the intermediate frequency IF2 (see equation (7.259)):

$$B_{\underset{L}{\text{U}}} \exp j\left\{(\pm\omega_{\underset{L}{\text{U}}} \pm \omega_{\text{LO1}} - \omega_{\text{LO2}})t \pm \underbrace{(\omega_{\text{LO1}} - \omega_{\underset{L}{\text{U}}})\tau_4 \pm \omega_{\underset{L}{\text{U}}}\tau_2}_{\omega\,\text{IF1}}\right\},\tag{7.261}$$

where ω_{IF1} (see equation (7.250)) reappears after the fourth parenthesis.

At $A6$: after the delay τ_5

$$A_{\underset{L}{\text{U}}} \exp\left\{j(\omega_{\text{IF2}}(t - \tau_5) - \omega_{\text{IF1}}\tau_3 \pm \omega_{\underset{L}{\text{U}}}\tau_1 \pm \Phi_1 + \Phi_2)\right\}\tag{7.262}$$

At $B6$: after the delay τ_6

$$B_{\underset{L}{\text{U}}} \exp\left\{j(\omega_{\text{IF2}}(t - \tau_6) - \omega_{\text{IF1}}\tau_4 \pm \omega_{\underset{L}{\text{U}}}\tau_2)\right\}.\tag{7.263}$$

After the crosscorrelation of the signals a (A_6) and b (B_6) from the two antenna elements and filtering out of all high frequency terms we get for USB and LSB

$$\langle a_{\underset{L}{\text{U}}}.b_{\underset{L}{\text{U}}}\rangle = A_{\underset{L}{\text{U}}} B_{\underset{L}{\text{U}}} \exp\left\{j(\omega_{\text{IF2}}(\tau_6 - \tau_5) + \omega_{\text{IF1}}(\tau_4 - \tau_3) \pm\right.$$
$$\left.\pm\, \omega_{\underset{L}{\text{U}}}(\tau_2 - \tau_1) \pm \Phi_1 + \Phi_2)\right\}.\tag{7.264}$$

The following are two alternative forms of the correlated outputs, for the USB and LSB:

$$= A_{\underset{L}{\text{U}}} B_{\underset{L}{\text{U}}} \exp\left\{j(\pm\omega_{\text{LO1}}(\tau_2 - \tau_1) + \omega_{\text{LO2}}(\tau_2 - \tau_1 + \tau_4 - \tau_3) +\right.$$
$$\left.+\, \omega_{\text{IF2}}(\tau_2 - \tau_1 + \tau_4 - \tau_3 + \tau_6 - \tau_5) \pm \Phi_1 + \Phi_2)\right\}.\tag{7.265}$$

This assumes that the second conversion is an upper sideband conversion.

If the second conversion is lower sideband, the following changes occur in equations (7.264) and (7.265): All terms originating before that conversion, that is containing ω_{IF1}, ω_{LO1}, ω_{LO2} (or ω_{U}, ω_{L}) change sign. This also applies to the two-phase rotation terms Φ_1 and Φ_2. Basically all terms coming from before the conversion considered change signs.

A third frequency conversion to ω_{IF3} will just add a term $\omega_{\text{IF3}}(\tau_8 - \tau_7) + \Phi_3$ to (7.264) in the case of an USB conversion. The third conversion will do the following to (7.265): replace the ω_{IF2} term by

$$\omega_{\text{LO3}}(\tau_2 - \tau_1 + \tau_4 - \tau_3 + \tau_6 - \tau_5) + \omega_{\text{IF3}}(\tau_2 - \tau_1 + \ldots - \tau_7) + \Phi_3\tag{7.266}$$

in case of USB conversion.

With the above remarks in mind, it is obvious that the expressions (7.264) and (7.265) have a general character. Only the first conversion, which may be double sideband, must be singled out. The other conversions can always be incorporated in terms of the form

$$\sum_i \omega_{\text{IF1}}(\tau_n - \tau_m) + \Phi_1'\tag{7.267}$$

in (7.264) or of the form

$$\sum_i \omega_{\text{LO}i}(\tau_2 - \tau_1 + \ldots) + \sum_i \Phi_i + \omega_{\text{IF}f}(\tau_2 - \tau_1 + \ldots),\tag{7.268}$$

where $\omega_{\text{LO}i}$ are successive LO frequencies, Φ_i' are the phase shifts in corresponding LO's and $\omega_{\text{IF}f}$ is the final (f) IF frequency, just before correlation. Having thus shown the general character of the expressions (7.264) and (7.265) we now only consider the case of *two* conversions in the following analysis (expressions: (7.264) and (7.265)).

As a radiosource moves across the sky, $(\tau_2 - \tau_1)$ is continuously changing which gives the interference fringes. Hence, if all delays and phase shifts are fixed in the system, only terms like $\omega_U(\tau_2 - \tau_1)$ or $\omega_L(\tau_2 - \tau_1)$ in (7.264) are varying with time. In the single side band (SSB) case, this shows that fringes are seen "at the observing frequency". By applying an appropriate variation to terms like Φ_1 or Φ_2 in (7.264) for U or L the output fringe frequency can be changed from the "natural" fringe rate to any desired rate. The terms varying with time like $\omega_U(\tau_2-\tau_1)$ or $\omega_L(\tau_2-\tau_1)$ are referred to as the interferometer phase.

7.10.2. Single Sideband (SSB) Signal Processing in a Finite Bandwidth

Suppose now that the receivers accept only a finite bandwidth. Then the equations (7.264) and (7.265) have to be replaced by integrals over the received bandpass. The expression (7.265) is easier to consider because it contains only *one* phase term which is a function of the frequency across the bandpass.

In practical cases one has $\omega_{IF2} < \omega_{LO2} < \omega_{LO1}$. Since $(\tau_2-\tau_1)$ has a given variation with time (set by diurnal motion of the tracked radio source), the terms like $\omega_{LO1}(\tau_2 - \tau_1)$ are varying faster than terms like $\omega_{IF2}(\tau_2 - \tau_1)$. Equation (7.265) contains "fast" terms of the type $\sum_i \omega_{LOi}(\tau_1 - \tau_2)$. In addition (7.265) contains "slow" terms of the kind $\omega_{IF2}(\tau_1 - \tau_2...)$

so that the output in a finite bandpass BP has the following form: USB/LSB

$$A'_{\substack{U \\ L}} = \exp\left\{j\left(\sum \omega_{LOk}(\tau_j - \tau_i) + \Phi'\right)\right\} \exp\left\{j\left(\omega_0(\tau_2 - \tau_1 + \right.\right.$$

$$\left.\left. + \tau_4 - \tau_3 + \tau_6 - \tau_5)\right)\right\}. \int_{BP} A_{\substack{U \\ L}} B_{\substack{U \\ L}} \exp\left\{j\left((-\omega_{IF2} - \omega_0).\right.\right.$$

$$\left.\left. .(\tau_2 - \tau_1 + \tau_4 - \tau_3 + \tau_6 - \tau_5)\right)\right\}.d\omega_{IF2} \, , \tag{7.269}$$

where ω_0 is the center frequency of the bandpass at the second IF and Φ' is a phase factor (corresponding to Φ'_i in (7.268)). From equation (7.269) it is seen that the fringes $\omega_U(\tau_2 - \tau_1)$ or $\omega_L(\tau_2 - \tau_1)$ are modulated by a term which is the Fourier transform of the received bandpass BP.

In other words, if all delays in the receivers are fixed, fringes are only observed near the "*white fringe*" condition (geometrical delay compensated by receiver delays). Around this situation, the fringe amplitude is enveloped by the Fourier transform of the receiver bandpass in the delay space (the so-called "sausage pattern"). In order to keep the fringe amplitude constant with time, one should permanently fulfill the condition

$$\tau_2 - \tau_1 + \tau_4 - \tau_3 + \tau_6 - \tau_5 = 0 \, , \tag{7.270}$$

so that the exp-function under the integral sign in (7.269) is maximum and does not vary in time.

In other words, one should constantly compensate the variation of the geometrical delay $(\tau_2 - \tau_1)$ as the source moves across the sky by one of the receiver delays $(\tau_4 - \tau_5)$ of $(\tau_6 - \tau_5)$. Physically this condition keeps the system close to the "white" fringe by retaining coherence between the various frequencies within the received bandpass. A look at the expression (7.265) shows that a slightly different situation occurs whether delay is tracked at the first or the second IF frequency. If the delay is continuously tracked at ω_{IF1} then the term $(\tau_2 - \tau_1) + (\tau_4 - \tau_3)$ is kept null and the fringe term is at frequency ω_{LO1}. If the delay is tracked at ω_{IF2}, the term $(\tau_2 - \tau_1) + (\tau_6 - \tau_5)$ is kept null and the fringe term is at $\omega_{LO1} \pm \omega_{LO2}$ depending on the relative signs of the two successive conversions.

Again the terms Φ_1 and/or Φ_2 allow to set the output fringe frequency to any desired value convenient for the detection system.

To improve the signal-to-noise-ratio (S/N) in the SSB case, one can make a second correlation (the so-called sine correlation as opposed to the previous "cosine" correlation).

This is generally done by introducing a phase shift across the bandpass in one antenna element before correlation. This multiplies the signal from that antenna by $\exp(j\pi/2)$ before multiplication in the correlator. This is equivalent to taking the imaginary part of equation (7.264) to (7.269). These "sine" fringes are in quadratic relative to the co-sine fringes. Because of the quadrature condition, the two noises at the ouput of the two correlators are statistically independent which indeed yields an improvement of $\sqrt{2}$ in signal-to-noise ratio relative to a single correlation. The cosine output is the real part of expressions (7.264) to (7.269), the sine output is the imaginary part.

7.10.3. Consequences of Delay Tracking in the SSB Finite Bandwidth Case

In general, delay is not tracked continuously but in discrete steps. These steps must be small enough so that the loss in amplitude (see equation (7.269)) due to the Fourier trans-form of the finite bandpass is kept below a specified limit (see: effective bandwidth and delay tracking in chapter 7.6). For a given allowed loss in amplitude, the broader the bandpass BP, the finer should be the delay step.

Between delay switchings, the fringes are at the observing frequency ω_U or ω_L. When delay is switched by $\Delta\tau$, a phase jump $\omega_{IF}.\Delta\tau$ occurs where ω_{IF} is the IF frequency at which the signal propagates through the switched delay line (see expr. (7.264)). In particular, in a multichannel receiver, the phase jumps are slightly different in different channels. Since delay switching does not occur at regular intervals, the best is to make those phase jumps as small as possible so they can be ignored by the reduction system. This often sets a more stringent limit on the smallest delay step than the amplitude loss condition mentioned above. If fringes are integrated over many delay switchings intervals and phase jumps are ignored, the reduction program works in the continuous delay trac-king approximation and the average fringe frequency is $\omega_{LO1} \pm \omega_{LO2}$. If phase jumps are too large to be ignored the reduction program has to process the data separately between two successive switching times. Between two such times, the fringe frequency is the signal frequency: ω_U or ω_L.

7.10.3.1. Double Sideband Processing

This is the case when both ω_U and ω_L are received at the RF steps and are superposed in the further IF stages. As mentioned earlier, because the two sidebands are uncorrelated, the correlated output voltages from them can simply be added. We shall also assume that $A_U B_U = A_L B_L = AB$ which is the strict DSB case. Then the real part of the sum of (7.265) yields

$$2AB \, \cos\Big(\Phi_1 + \omega_{LO1}(\tau_2 - \tau_1)\Big) \, \cos\Big(\Phi_2 + \omega_{LO2}(\tau_2 - \tau_1 + \tau_4 - \tau_3) +$$

$$+ \, \omega_{IF2}(\tau_2 - \tau_1 + \tau_4 - \tau_3 + \tau_6 - \tau_5)\Big) . \tag{7.271}$$

The imaginary part of the sum of the parts of equation (7.264) yields

$$2AB \, \cos\Big(\Phi_3 + \omega_{LO1}(\tau_2 - \tau_1)\Big) \, \sin\Big(\Phi_2 + \omega_{LO2}(\tau_2 - \tau_1 + \tau_4 - \tau_3) +$$

$$+ \, \omega_{IF2}(\tau_2 - \tau_1 + \tau_4 - \tau_3 + \tau_6 - \tau_5)\Big) . \tag{7.272}$$

The real part is the output of the cosine correlator, the imaginary part is the output of the sine correlator.

In the finite bandwidth case, the expressions (7.271) and (7.272) are further multiplied by the Fourier transform of the bandpass if the response of the system is the same in both antennas (or of the product of the voltage responses of the two arms of the interferometer). These last expressions show the following: Fringes occur at ω_{LO1} whether or not delay is tracked. The phase of the fringes is independent of the receiver delays $\tau_3, \tau_4, \ldots$. This physically comes from the fact that the phase shifts introduced by receiver delays act with opposite signs on either input sideband. The fringe amplitude is multiplied by a slower term (varying as $(\omega_{LO2} + \omega_{IF2}).(\tau_2 - \tau_1)$) if receiver delays and phase shifts are fixed. Again, if one wants constant (and maximum) fringe amplitude one should track delay, that is compensate the geometrical delay variation by a receiver delay variation. Thus by tracking delay the argument of the second cosine in (7.272) is zero and no fringes are seen at the output of the sine correlator. This means that the factor of 2 improvement of the double sideband mode for continuum sources turns into a $\sqrt{2}$ only because the sine correlator becomes useless. Setting Φ_2 such that both sine and cosine correlators give equal amplitude outputs (reduced by $\sqrt{2}$ relative to the maximum) will now improve the situation because two outputs of amplitude $1/\sqrt{2}$ are equivalent in S/N ratio to one output of amplitude unity (with statistically independent noises).

Similarly to the SSB case, the goal of the delay tracking is to keep coherence between all frequencies within the input bandpass, the "white fringe" condition. In the DSB case, the bandpass is made of two separate sidebands. The condition on the smallest delay step for a specified amplitude loss is thus generally more stringent. The delay (or "sausage") pattern is given by the second cosine in (7.271). In the practical case of finite bandwidth, this delay pattern is further multiplied by the delay pattern of the SSB case, that is the Fourier transform of the individual sidebands. We now examine the different ways of tracking delay and their consequences.

If delay is tracked at the first IF, one assumes $\Phi_2 = 0$, $\tau_5 = \tau_6$ and one constantly tracks $\tau_4 - \tau_3$ so that: $\tau_2 - \tau_1 = \tau_4 - \tau_3$. In this case, between delay switchings, the fringe amplitude behaves (see (7.271)) as:

$$\cos(\omega_{LO2} + \omega_{IF2})\Delta\tau = \cos(\omega_{IF2}.\Delta\tau) , \qquad (7.273)$$

where $\Delta\tau$ is the delay error

$$\Delta\tau = (\tau_2 - \tau_1) - (\tau_4 - \tau_3) . \qquad (7.274)$$

If the delay is tracked at the second IF, one tracks $\tau_6 - \tau_5$ so that

$$\tau_2 - \tau_1 = \tau_6 - \tau_5 . \qquad (7.275)$$

This condition, which is sufficient in the SSB case (in each individual bandpass) is not sufficient to keep the argument of the second cosine of (7.271) equal zero. In addition to $\tau_3 = \tau_4$ one should also track Φ_2 as the source moves across the sky so that

$$\Phi_2 = \omega_{LO2}(\tau_2 - \tau_1) . \qquad (7.276)$$

If Φ_2 and the delay $(\tau_6 - \tau_5)$ are switched in discrete steps, between switching of Φ_2 or $(\tau_6 - \tau_5)$ the fringe amplitude still behaves as

$$\cos\{(\omega_{LO2} + \omega_{IF2})\Delta\tau\} = \cos(\omega_{IF1}.\Delta\tau) . \qquad (7.277)$$

If, however, switching of Φ_2 is done in much smaller steps than the steps in $\omega_{\text{IF2}}(\tau_6 - \tau_5)$, the fringe amplitude between delay switching behaves as $\cos(\omega_{\text{IF2}}.\Delta\tau)$. Since $\omega_{\text{IF2}} < \omega_{\text{IF1}}$, this makes less stringent the requirement on the smallest delay step to meet a specified loss in fringe amplitude.

In summary, if delay is not tracked at the first IF, a phase rotator *must* exist in the second LO.

One advantage of tracking delay at an IF lower than the first IF frequency is to relax the requirement on the smallest delay step for a specified phase jump in the SSB case (see chapter 7.6.3).

7.10.4. Possible Ways of Separating the Sidebands in case of DSB Reception

A given type of astronomical observation may be optimum in SSB or DSB reception. It is therefore desirable to easily switch from one mode to the other.

At millimeter wavelength, where the input (RF) stage of the receiver is a mixer, reception is most generally done in DSB mode. Rejection of the unwanted sideband for a particular application is better done with a filter placed at the input of the receiver before the mixer. In the case of an interferometer this requires a mechanical input device with very stringent mechanical specifications so that no differential phase shift occurs between two antenna elements. To bypass this added difficulty, it is more convenient to separate the sidebands within the receiver and this is possible by using phasing and/or fringe frequency separation as we now will show:

7.10.4.1. Sideband Separation by Phasing

Simply adding the complex signals of the U and L parts of (7.264) or (7.265) at the correlator ouputs, gives

$$A_U B_U \exp\{j(\psi_U + \Phi_1 + \Phi_2)\} + A_L B_L \exp\{j(\psi_L - \Phi_1 + \Phi_2)\} . \tag{7.278}$$

This shows that the two phase shifts Φ_1 and Φ_2 have different actions on the phase of the sidebands.

The following sequence of values of Φ_1 and Φ_2 will give a correlated output equal to the real part of (7.278), namely

Φ_1	Φ_2	Correlated output
0	0	$+A_U B_U \cos\psi_U + A_L B_L \cos\psi_L$
$+\dfrac{\pi}{2}$	$-\dfrac{\pi}{2}$	$+A_U B_U \cos\psi_U - A_L B_L \cos\psi_L$
0	$+\dfrac{\pi}{2}$	$-A_U B_U \sin\psi_U - A_L B_L \sin\psi_L$
$+\dfrac{\pi}{2}$	0	$-A_U B_U \sin\psi_U + A_L B_L \sin\psi_L$

Table 7.3. Correlated output by sequences of Φ_1, Φ_2.

It is assumed that these offsets on Φ_1 and Φ_2 are applied on top of the time variation on Φ_1 and Φ_2 which determine the chosen fringe rate. After this phasing sequence of both phase shifters, the computer can solve for $A_U B_U$ and $A_L B_L$, the amplitudes in the upper and lower sidebands and for the phases ψ_U and ψ_L. It is also interesting to note that this phasing method (which is an application of the principle of sideband separation mixers) allows to solve another difficulty, namely the impossibility of phase measurement when the output fringe frequency is zero in the DSB case. In this case, only one correlator is available (see above) and a non-zero fringe rate is necessary to measure the fringe phase whereas in the SSB case, the ratio of sine and cosine outputs provide the phase. However, with the phasing sequence shown above, it is possible to measure a phase in all circumstances. For the strict DSB case ($A_U B_U = A_L B_L = AB$), it is easy to see from (7.271) that the first and the fourth states of the phasing sequence provides the cosine and the sine of the fringe phase.

The first state provides: $\cos \omega_{LO1}(\tau_2 - \tau_1) \cos(\Phi_2 + ...)$
the fourth state provides $- \sin \omega_{LO1}(\tau_2 - \tau_1) \cos(\Phi_2 + ...)$,
where the argument of the second cosine does not change.

7.10.4.2. Sideband Separation by Fringe Rate

The natural fringe rates of the two sidebands are different by a fractional amount equal to $2\omega_{IF}/\omega_{LO}$. This, in principle, could allow sideband separation at the output by numerical filtering. This fringe rate difference, however, is too small to achieve proper selectivity in all cases. In particular, separation is practically impossible, when the natural fringe rate is small. In this case, the required selectivity (typically: 20.dB) would call for exceedingly long integration times. The use of the two phase rotators Φ_1 and Φ_2 solves this difficulty because the output fringe rates of the two sidebands can be set to any desired rate. Indeed, if $\Phi_1 = \Phi_1' + \Omega_1 t$ and $\Phi_2 = \Phi_2' + \Omega_1 t$, expressions (7.264) show that the fringe phase of the upper sideband is

$$\omega_U(\tau_2 - \tau_1) + \Phi_1' + \Phi_2' + (\Omega_1 + \Omega_2)t$$

while the fringe phase of the lower sideband is

$$-\omega_L(\tau_2 - \tau_1) - \Phi_1' + \Phi_2' + (\Omega_1 + \Omega_2)t .$$

In other words, the fringe rates of the two sidebands are

$$\omega_U \dot{\tau} + \Omega_2 + \Omega_1$$

and

$$-\omega_L \dot{\tau} + \Omega_2 - \Omega_1 ,$$

where

$$\dot{\tau} = \frac{d}{dt}(\tau_2 - \tau_1) \tag{7.279}$$

is the first derivation of the delay related to the time. The two phase rotators $\Phi_{1,2}$ can be set to appropriate values Ω_1 and Ω_2 so that the fringe frequencies are easily separated.

7.10.4.3. Concluding Remark on Sideband Separation

The two methods of sideband separation outlined above have a disadvantage in that they do not eliminate the noise of the unwanted sideband. An improvement in signal-to-noise ratio is often obtained if the mixer "sees" a cold load in the unwanted sideband. This is an advantage of sideband rejection at the RF stage with the added mechanical constraints to preserve *phase stability*. A way to relax the mechanical specifications of an RF filter is to inject the LO through the same system; in this case the mechanical stability requirement is relaxed by the ratio ω_{IF}/ω_{LO}.

Summarizing, a variable delay system must constantly keep the system close to the "white fringe" in order to ensure coherence over the entire band of received frequencies. A small delay error yields a phase error and a slight amplitude loss in the single sideband (SSB) case. A delay error does not effect the phase in the double sideband (DSB) case but it yields an amplitude loss. The variation of output fringe amplitude as a function of delay offset from the "white fringe" condition is the Fourier transform of the input bandwidth (i.e., the product of voltage responses of the two arms).

In a DSB system, if the variable delay is not at the first intermediate frequency, a second phase rotator is needed in order to keep amplitude coherence. In the SSB case, the delay tracking accuracy is determined by the input bandwidth. In the DSB case, the accuracy is determined again by the "effective" input bandwidth (Δf_{eff}) that is twice the first IF ($\Delta f_{eff} = 2f_{IF1}$). In the double sideband case, if the variable delay is operating at an IF *other* than the first, and if the second phase rotator inserted in one of the LO's is tracking continuously relative to the variable delay, the accuracy of delay tracking is determined by the IF which has the variable delay. Finally, the avoidance of phase jumps in the SSB case also sets how fine delay tracking should be.

Appropriate adjustment of the phases and/or frequencies of two phase rotators in the first LO and one of the following ones allows separation of signals from the two received sidebands without the use of front end filters. This technique is not optimum in terms of signal-to-noise ratio S/N.

Two crosscorrelators in quadrature improve the signal-to-noise ratio by $\sqrt{2}$ in the SSB case over the use of one correlator. In the DSB case, only one correlator is meaningful so that the improvement is only $\sqrt{2}$ over the SSB case for continuum observations.

Various offsets, drifts and spurious responses can be eliminated by introducing a phase switch (between 0 and π) as early as possible in the system, that is in the first local oscillator.

The Fourier transform relationship between fringe amplitude (and phase) and input bandpass is the basis of the principle of *spectral correlators*.

The *"delay beam"* *effect* is only important for VLBI. Until this point it has been assumed that a point source is tracked at the center of the primary beam of the antenna elements, that is at the center of the "field of view". The difference $\tau_2 - \tau_1$ was the geometrical delay for the field center. However, this geometrical delay varies across the field of view. If the antenna spacing is large enough the geometrical delay may vary significantly within the field of view in such a way that significant loss in amplitude (due to the loss of coherence) occurs for sources at the edge of the primary beam. If the delay is set right for the center of field, the loss of coherence away from the field center brings an attenuation which adds to the attenuation of the primary beam. This second beam is referred to as the "delay beam". For large enough spacings, the variation of delay across the sky is so fast that the delay beam becomes narrower than the main beam of the antenna elements. In this case, it is the delay beam and not the primary beam which determines at *actual field of view*.

As an example consider the delay beam of a mm-wave-interferometer: the effect is most severe in the double sideband mode for continuum observations. In this case, the first zero of the delay pattern $\cos 2f_{IF1}\Delta\tau$ occurs at $\Delta\tau = 1/4 f_{IF1}$. The *actual* delay is $\tau = (D/c)\cos\vartheta$, where $\vec{D}$ is the baseline vector length, c is the velocity of light and ϑ is the angle between the radiosource direction and the baseline vector. In the worst case (*EW baseline*, source at transit): $d\tau = (D/c)d\vartheta$.

The first zero of the antenna beam occurs at $d\vartheta = \lambda/D_p$, where λ is the operating wavelength and D_p is the (parabolic reflector) antenna element diameter. Hence, the delay beam effect does not occur if the delay error at the edge of the beam is much smaller than

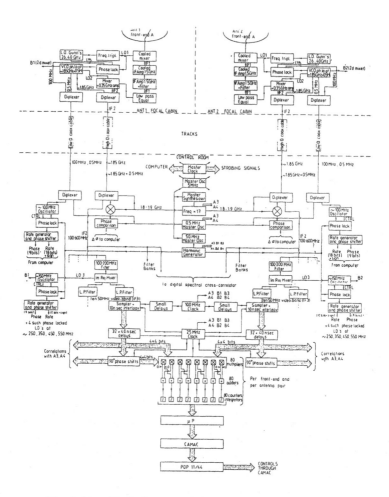

Fig. 7.54. Block diagram of the electronics for a mm-wavelength twin interferometer and one front-end per antenna (256,257).

the first zero of the delay pattern, that is

$$(B/c)(\lambda/D_p) \ll 1/(4f_{\mathrm{IF1}}) . \tag{7.280}$$

Using the typical parameter for a mm-wavelength-interferometer:

$f_{\mathrm{IF}} = 1.5$ GHz, $\lambda = 3.$mm, $D_p = 15.$m yields: $D \ll 250.$m .

This means that for continuum DSB observations, the delay beam effect will be severe. Hence, the methods of sideband separation outlined in the following will be useful.

In the SSB case, the delay pattern is $\text{si}(\pi\Delta f\Delta\tau)$, where Δf is the bandwidth: the first zero of the delay pattern is at $\Delta\tau = 1/\Delta f$. Then the above delay beam effect condition (7.280) becomes

$$(D/c)(\lambda/D_p) \ll 1/\Delta f \ .$$

With $\Delta f = 500.\text{MHz}$ this yields: $D \ll 3.\text{km}$ (!).

It is worth remarking that, even if the delay beam effect occurs, it does not play a role for a point source at the center of the beam which is the case when phase calibrators are observed.

This problem is particularly severe for DSB continuum observations for which the delay beam is narrow due to the very large effective bandpass (or twice the first IF frequency). If, on the other hand, a double side band (DSB) receiver is used to simultaneously observe a line in each sideband, this problem does not occur provided the two lines appear in separate frequency channels at the output of the receiver back-end. In this case, the superposition of both sidebands represented by expression (7.271) does not occur at the same output correlator.

Expression (7.271) also shows that the delay beam is independent of the way delay is tracked. Indeed, if $\Delta(\tau_2 - \tau_1)$ is the variation of geometrical delay across the sky, the second cosine term in (7.271) becomes

$$\cos\left\{(\omega_{LO2} + \omega_{IF2})\,\Delta(\tau_2 - \tau_1)\right\} = \cos\omega_{IF1}\,\Delta(\tau_2 - \tau_1)\ ,$$

instead of 1 at the field center and this is true whatever the delay tracking scheme (1st or 2nd IF frequency) is.

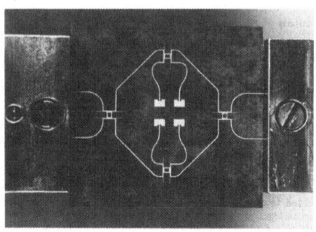

Phase comparison monopulse quasi-optical system (vert. and hor. plane).

Antenna elements:	Four patches (monolithic technique, in the focal plane of a lens)
Monopulse comparator(s):	Four stripline ring hybrids
Substrate:	RT/Duroid, t = 0.124 mm
Center frequency (RF):	94. GHz (λ_0 - 3.19 mm)
Bandwidth (Δf):	0.5 GHz
Scale:	4 to 1

(Courtesy of Dipl.-Ing. A. Schlaud, Telefunken Systemtechik GmbH, D-7900 Ulm 1, Fed. Rep. of Germany)

8. RADAR APPLICATIONS OF INTERFEROMETRY

8.1. Introduction

There are not only applications of interferometry for radio astronomy but also for radar techniques. Particularly the increasingly important field of space research requires methods which yield measurements of position and velocity of flying objects to the highest precision. The interferometer used in radio astronomy is a passive one, as the energy source of emitted radiation is identical with the target. A tracking radar which uses *phase information* can however be regarded as an active interferometer and referred to as an interferometer radar; the expressions *simultaneous phase-comparison radar* and phase comparison *monopulse* radar are also in use (76,77,206).

There are two categories of measuring methods:

a) *tracking:* a ground station is used to follow the motion of a flying object,

b) airborne radar: the radar is on board of the flying object (78).

8.2. Fundamentals

The range R to the target is so large compared to the distance D (baseline) between the two antenna elements that the two incident rays can be assumed parallel. The distance of antenna element A from the target is (Fig. 8.1)

$$R_1 = R + (D/2) \sin \vartheta \tag{8.1}$$

with the zenith angle ϑ and the distance between the target and element D is

$$R_2 = R - (D/2) \sin \vartheta . \tag{8.2}$$

Following equation (6.1) the phase difference between these signals is

$$\Delta \Phi = (2\pi D/\lambda) \sin \vartheta . \tag{8.2a}$$

$2k\pi$ must be added to this value since the number of additional 360.° cycles of phase is unknown (*phase ambiguity*). The significant advantage of this procedure is that the flying object only requires a *transponder*, which is interrogated by the ground station. The measuring accuracy is very high: A transmitted signal of 3. m wavelength allows a phase resolution corresponding to 3. mm. For $D = 150.$ m the figure of 3. mm corresponds to 20 μ Rad or 4″ of arc. The highest accuracy is achieved perpendicular to the baseline. For a range to a spacecraft of one astronomical unit of 150. million km the error is about $\Delta R = 3000.$ km and $\Delta R = 8.$ km for a spacecraft near the moon.

The angular resolution of the interferometer is maximum at right angles to the baseline, i.e. at an angle of incidence $\vartheta = 0°$ and decreases with the cosine of ϑ for smaller angles.

Tropospherical refraction worsens the resolution. In order to get higher accuracy the baseline must be increased. However, this introduces new problems, since simultaneous direct phase comparison of the received signals is difficult with large baselines. Further, the direction is no longer a simple function of the range difference because the incident rays cannot be considered parallel. Instead of range evaluation via direct phase comparison the ranges between the flying object and each interferometer element must be found before evaluation of the *direction parameters* is possible. The sine of the angle of incidence is then equal to the difference of the individual ranges divided by the baseline length. If an improvement of the angular resolution by increasing the baseline is required, the range and *phase errors* must increase less fast than baseline length. The enlargement in baseline from 150. m to 1500. km for the same angular accuracy of 20 μ rad. requires a baseline accuracy of $\Delta D = 30.$ m. Such accuracies of station coordinates on earth require a high geodetic measuring accuracy which again requires an accurate knowledge of the size and figure of the earth.

At the time the US *Minitrack* interferometric network (77) was developed, the orbits of satellites were planned as circles. As space technology progressed the orbits became more elliptical, particularly for planetary missions. The interferometric method proved inadequate for measurements at apogee of spacecraft in such orbits (24). For this special case the "Goddard Range and Rate Radar" was developed, which will not be dealt with here (207).

8.2.1. Doppler Shift

It is well known in acoustics and optics that rapid motion of a source of radiation or of an observer give rise to an apparent frequency shift. This is the Doppler effect (10,78). The shifted frequency coming from a moving source can be evaluated as follows (77)

$$f = \{1 \pm (v/c)\} f_0 = f_0 \pm f_0(v/c) = f_0 + (v/\lambda) = f_0 \pm f_d , \qquad (8.3)$$

where

f_0 = transmitted frequency
v = radial velocity of the transmitted source relative
      to observer
c = velocity of light (or sound)
λ = free space wavelength.

The term f_d expresses the shift caused by the Doppler effect. As f or f_d can be determined by measurement (index m), the *velocity* of the *target* relative to the observer may be found.

$$v = \pm(f_m - f_0)\lambda = \pm f_d\lambda . \qquad (8.4)$$

In the case of considerable baseline lengths (D) encountered in interferometers the knowledge of the difference of the Doppler shift between the single elements is of interest.
Fig. 8.1 (top) shows that this Doppler difference is given by the magnitude of the differential variation of the path length divided by the wavelength λ

$$\Delta f_d = \Theta(D/\lambda) \cos \vartheta = (v \sin \alpha/R)(D \cos \vartheta/\lambda) , \qquad (8.5)$$

where Δf_d is the difference of the Doppler shifts for the two elements. Here

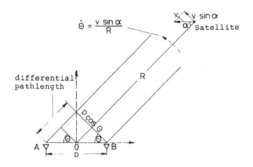

Fig. 8.1.   To the Doppler effect of
             fastly moving targets.

R = range of the target
α = angle between the velocity vector of the target and the
     direction to the origin.

Δf_d is at a maximum when the target moves exactly over the array, i.e., $\alpha = 90.°$ and
$\vartheta = 0.°$ and is given by

$$\Delta f_d^{\max} = vD/(R \cdot \lambda) \,. \tag{8.6}$$

Consider a satellite with a velocity $v = 8000.$ m/s at a height of 556. km flying across
an interferometer of $D = 31.$ m producing a *Doppler shift* of $f_d = 16.$ Hz at $\lambda = 3.$ m
wavelength.

8.2.2. Defocussing Effects Produced by Target Movement

The fact that certain targets, for example satellites, move with high velocity leads to the
step, that also the *(transit) time* of the signal coming from a highly directive antenna (26)
must be considered. The following principal effects are found:

a. The pencil beam from the transmitter reaches a certain position in space which the
   target had occupied at the time the signal was transmitted, so that the ray tracking
   the target is always somewhat *behind* the target position.

b. In general the energy is not focussed at the location of the target, not even at the
   delayed position discussed above, because at some particular time the phases at the
   i-th and k-th interferometer elements correspond to different target positions.

Let us consider the following example:
A target moves with the tangential velocity v_t parallel to the baseline vector $\vec{D}$ between
the twin interferometer elements. This baseline has the length $|\vec{d}|$ and the target transmits
a pulse every D/c seconds. When the more distant element from the target receives the
first pulse, then the nearer element has already received the second pulse unless the target
has moved only a negligible fraction of the wavelength from its previous position between
the two pulses. The first error a. caused by wrong *beam pointing* can be neglected in this

example and the expected *phase error* is proportional to the distance moved during one *pulse interval*: $v_t, D/c$ (Fig. 8.2).

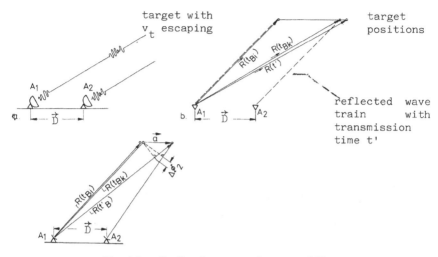

Fig. 8.2.   *To the phase error by target shift*
   a. measuring equipment
   b. target coordinates after shifting
   c. phase difference.

If the respective target distances are $\vec{R}(t_1)$, $\vec{R}(t_2)$, the phase difference is

$$\Phi = (2\pi/\lambda)\{|\vec{R}(t_1) - \vec{R}(t_2)|\} = (2\pi/\lambda)v_t(D/c) \qquad (8.7)$$

and the phases at the elements

$$\Phi_i = (2\pi/\lambda)|\vec{R}(t_1)| \qquad (8.8)$$

$$\Phi_k = (2\pi/\lambda)|\vec{R}(t_2)| \ . \qquad (8.9)$$

$\vec{R}(t)$ can be defined as the position vector of the target in relation to the i-th element (Fig. 8.2) and $\dot{\vec{R}}(t)$ as the first derivative, $\vec{D}$ is the vectorial shift of the i-th element from the k-th element. We consider the "wave packet" which returns to the i-th element at time t' after reflection from the target at an earlier time t_{Bi}, i.e. at the time when the target had the position $\vec{R}(t_{Bi})$. The following equations represent a useful approximation for the time t_{Bi} and the position $\vec{R}(t_{Bi})$ at which the reflection took place (17)

$$t_{Bi} \approx t' - |\vec{R}(t')|/c \qquad (8.10)$$

$$\vec{R}(t_{Bi}) \approx \vec{R}(t') - (t' - t_{Bi})\dot{\vec{R}}(t') = \vec{R}(t') - \frac{|\vec{R}(t')|\dot{\vec{R}}(t')}{c} \ . \qquad (8.11)$$

At the time t' another wave, which was, however, not reflected at the same time by the target as the i-th wave, returns to the k-th element, because it had to traverse a different distance. The reflection time and location of this wave packet t_{Bk} and $\vec{R}(t_{Bk})$ are defined by the following equations:

$$t_{Bk} = t' - \{|\vec{R}(t') - \vec{D}|\}/c \tag{8.12}$$

$$\vec{R}(t_{Bk}) = \vec{R}(t') - \{|\vec{R}(t') - \vec{D}|\dot{\vec{R}}(t')\}/c . \tag{8.13}$$

As the phase of a reflected wave packet depends only on the phase of the source and the path length, a focussing at $\vec{R}(t_{Bi})$ would result if both returned waves had been reflected at this position.

The k-th wave was reflected at another point, so that an additional correcting phase φ_1 must be introduced in order to get a focussing at $\vec{R}(t_{Bi})$. φ_1 compensates the additional path length which the k-th wave must traverse until it reaches the target. After reflection the wave travels the same path length a second time. The following equations describe the situation:

$$\Phi_1 = 2\frac{\omega_0}{c}\{|\vec{R}(t_{Bk})| - |\vec{R}(t_{Bi})|\} =$$

$$= 2\frac{\omega_0}{c}\left\{\left|\vec{R}(t') - \frac{|\vec{R}(t') - \vec{D}|}{c}\dot{\vec{R}}(t')\right| - \left|\vec{R}(t') - \frac{|\vec{R}(t')|\dot{\vec{R}}(t')}{c}\right|\right\} =$$

$$= 2\frac{\omega_0}{c}\left\{\left|\frac{|\vec{R}(t')|}{\vec{R}(t')}\left(\frac{\vec{R}(t')\vec{R}(t')}{|\vec{R}(t')|} - \frac{\vec{R}(t')\vec{R}(t')}{|\vec{R}(t')|c}|\vec{R}(t') - \vec{D}|\right)\right| -\right.$$

$$\left. - \left|\frac{|\vec{R}(t')|}{\vec{R}(t')}\left(\frac{\vec{R}(t')\vec{R}(t')}{|\vec{R}(t')|} - \frac{\vec{R}(t')|\vec{R}(t')|}{|\vec{R}(t')|c}\dot{\vec{R}}(t')\right)\right|\right\} =$$

$$= 2\frac{\omega_0}{c}\left\{\left[|\vec{R}(t')| - \frac{\vec{R}(t')\dot{\vec{R}}(t')}{c}\frac{|\vec{R}(t') - \vec{D}|}{|\vec{R}(t')|}\right] - \left[|\vec{R}(t')| - \vec{R}(t')\dot{\vec{R}}(t')\frac{1}{c}\right]\right\} =$$

$$= 2\frac{\omega_0}{c^2}\left\{\vec{R}(t')\dot{\vec{R}}(t')\left[1 - \frac{|\vec{R}(t') - \vec{D}|}{|\vec{R}(t')|}\right]\right\} =$$

$$= 2\frac{\omega_0}{c^2}\left\{\vec{R}(t')\dot{\vec{R}}(t')\left[1 - \left|\frac{|\vec{R}(t')|}{\vec{R}(t')|\vec{R}(t')|}\left(\frac{\vec{R}(t')\vec{R}(t')}{|\vec{R}(t')|} - \frac{\vec{R}(t')}{\vec{R}(t')}\vec{D}\right)\right|\right]\right\} =$$

$$= 2\frac{\omega_0}{c^2}\left\{\vec{R}(t')\dot{\vec{R}}(t')\left[1 - \frac{1}{|\vec{R}(t')|}\left(|\vec{R}(t')| - \frac{\vec{R}(t')}{|\vec{R}(t')|}\vec{D}\right)\right]\right\} =$$

$$= 2\frac{\omega_0}{c^2}\left\{\vec{R}(t')\dot{\vec{R}}(t') - \vec{R}(t')\dot{\vec{R}}(t') + \frac{[\vec{R}(t')\dot{\vec{R}}(t')][\vec{R}(t')\vec{D}]}{|\vec{R}(t')|^2}\right\}$$

$$\Phi_1 = 2\frac{\omega_0}{c^2}\frac{[\vec{R}(t')\dot{\vec{R}}(t')][\vec{R}(t')\vec{D}]}{|\vec{R}(t')|^2} . \tag{8.14}$$

If Φ_1 is added to a, the returning waves at the point $\vec{R}(t_{Bi})$ are in phase at the time t'_B. At the time at which they arrive there the target has moved again to the new position $\vec{R}(t'_B)$. t'_B and $\vec{R}(t'_B)$ are approximately given by

$$t'_B = t_{Bi} + (2/c)\{|\vec{R}(t')|\} \tag{8.15}$$

$$\vec{R}(t'_B) \approx \vec{R}(t_{Bi}) + (t_{Bi} - t'_B)\,\dot{\vec{R}}(t')$$

$$\vec{R}(t'_B) = \vec{R}(t_{Bi}) + (2/c)\,|\vec{R}(t')|\,\dot{\vec{R}}(t') \ . \tag{8.16}$$

The vector $\vec{a}$ gives the position of the target relative to the focussing point (Fig. 7.33)

$$\vec{a} = \vec{R}(t'_B) - \vec{R}(t_{Bi}) = (2/c)\,|\vec{R}(t')|\,\dot{\vec{R}}(t') \ . \tag{8.17}$$

In order to actually get a focussing at the position of the target, a further correction Φ_2 relative to the phase of the i-th element must be added.

This correction can be visualised with the aid of Fig. 8.2c and is approximately

$$\Phi_2 \approx \frac{\omega_0}{c}\left\{\frac{\vec{a}\,\vec{R}(t')}{|\vec{R}(t')|} - \frac{\vec{a}\,[\vec{R}(t') - \vec{D}]}{|\vec{R}(t') - \vec{D}|}\right\} \approx$$

$$\approx \frac{\omega_0}{c}\ \frac{1}{|\vec{R}(t')|\,|\vec{R}(t') - \vec{D}|}$$

$$\left\{\left[|\vec{R}(t') - \frac{\vec{R}(t')\vec{D}}{\vec{R}(t')}\right]\vec{a}\,\vec{R}(t') - |\vec{R}(t')|\,\vec{a}\left[\vec{R}(t') - \vec{D}\right]\right\} \approx$$

$$\approx \frac{\omega_0}{c\,|\vec{R}(t')|^2}\left\{|\vec{R}(t')|\,\vec{a}\,\vec{D} - \frac{[\vec{R}(t')\vec{D}]\,[\vec{a}\,\vec{R}(t')]}{|\vec{R}(t')|}\right\}$$

$$\Phi_2 \approx \frac{2\omega_0}{c^2}\,\vec{D}\,\dot{\vec{R}}(t') - \frac{2\omega_0}{c^2}\,\frac{[\vec{R}(t')\vec{D}]\,[\vec{R}(t')\dot{\vec{R}}(t')]}{|\vec{R}(t')|^2} \ . \tag{8.18}$$

By addition of Φ_1 and Φ_2 we obtain the total *correcting phase* necessary in order to keep the beam of a self-focussing system on target

$$\Phi = \Phi_1 + \Phi_2 = \frac{2\omega_0}{c^2} \cdot \dot{\vec{R}}(t')\vec{D} \ . \tag{8.19}$$

Consequently, the sum of the errors described above is directly proportional to the baseline length (distance between elements) and the velocity of the target. The maximum value of the phase correction appears when $\vec{R}(t')$ and $\vec{D}$ are maximum. Two typical maximum values of these parameters are

$$|\vec{D}| = 3.\text{ km}, \quad |\dot{\vec{R}}(t')| = 6,6 \text{ km/s} \ .$$

This means e.g. for an X-band radar system (8.2–12.4 GHz) that the phase error must be less than $15.°$ in order to achieve adequate ray focussing.

8.2.3. Resolution of the Angular Ambiguity

As was known in chapter 5.6 (Fig. 6.2) the number of sidelobes of the same amplitude as the mainlobe (grating lobes) increases with growing baseline length. Parallel to this effect the half power beamwidths of these lobes decrease, which corresponds to an increase of angular resolution. As pointed out, one has to pay for the higher angular resolution with *ambiguities*, as, for target positions, away from the zenith, one cannot know which of the grating lobes has seen the target; i.e. the phase difference φ measured with the interferometer corresponds to many possible target positions. It is possible to alleviate this problem by using a third element which is on the same baseline but with a smaller baseline length in relation to one of the original elements. The result is a double interferometer with different baseline lengths, of which the short-baseline array gives a coarse but unambiguous localisation and the second, longer-baseline interferometer produces an accurate angle measurement (143). Because of their high angular resolution interferometric methods have also found favour in the field of radar. However, in addition to the well-known quasioptical definition of angular resolution (4) derived from the antenna geometry and pattern parameters another, more general definition must be considered: the angular resolution of a physical device is very generally defined as the minimum distance of two input signals, from which the resultant output signal has just two peaks with a 3dB drop between them. This qualitative definition says nothing about the influence of the *form* of the *signal* of the noise; even though the angular resolution depends strongly on the *form* of the *signal*.

The interrelation between angular resolution and form of the received signal is described by the *ambiguity function*. This is the *crosscorrelation* function of a copy of the transmitted signal and the doppler-shifted echo. In the special case of a Doppler frequency shift $f_d = 0$ the ambiguity function is identical to the *auto-correlation function (acf)* of the transmitted signal. The desired shape of the magnitude of the ambiguity function can be formulated after having defined a radar problem; but it cannot be used to calculate the appropriate signal form because the knowledge of *ambiguity phase* is missing. The *ambiguity function (a.f.)* represents a means of investigating the angular resolution of a given signal form.

As the multilobe structure of an interferometer beam can give rise to *ambiguities*, it is necessary to investigate the influence of this phenomenon on the ambiguity function. In modern radars modulated signals are normally used and it can be shown that by choice of a suitable signal bandwidth ambiguities can be avoided. It is possible to lower the sidelobe amplitude to the $1/N$ part of its normal spectral magnitude, where N is the number of elements in a long signal sequence. It is also necessary to investigate what influence atmospheric refraction has on angle measurement by an interferometer, which is of special interest for *tracking* satellites. Finally, some interferometer applications are discussed, such as an interferometer using a pseudo-random coded signal.

With an interferometer system the angle of incidence ϑ can be determined by measuring the phase difference Φ of signals received by two antennas separated by a baseline length D (see equation (6.1))

$$\Phi = (2\pi/\lambda)D \sin \theta \ . \tag{8.20}$$

First the operation of an interferometer will be treated in the one-dimensional case. The computation of the antenna pattern of a linear *antenna array* of N equal elements results from the *multiplicative law* of superposition (3,32,50) so that for a one-dimensional case we can write

$$C(\vartheta) = \sum_{n=0}^{N-1} \underline{L}(\vartheta)\, p_n\, e^{-j\Phi_n} = \underline{L}(\vartheta)\, \underline{M}(\vartheta) \ , \tag{8.21}$$

where

$\underline{L}(\vartheta)$ is the complex vertical beam of an antenna

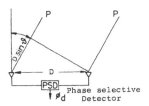

Fig. 8.3.   Geometry of the twin interferometer.

p_n is the power excitation of element n,
Φ_n is the phase difference between elements n and $n-1$,
$\underline{M}(\vartheta)$ is the complex array factor (for angle ϑ only).

For a constant excitation over the array

$$p_n = p_0, p_1, ..., = 1 \tag{8.22}$$

the equation for the array factor of N elements is

$$\underline{M}(\vartheta) = \sum_{n=1}^{N-1} e^{-j\Phi_n} = \frac{\sin(N\Phi/2)}{\sin(\Phi/2)} \ , \tag{8.23}$$

where Φ_n is defined by equation (8.20). With inter-element distance (baseline) normalized to the free space wavelength λ

$$k = D/\lambda \tag{8.24}$$

and $N = 2$ (8.23) may be written

$$\underline{M}(\vartheta) = \{\sin(2\pi k \sin \vartheta)\}/\sin(\pi k \sin \vartheta) \ . \tag{8.25}$$

Fig. 8.4 is a representation of this relation of isotropic radiators (see equation (8.22)) polar coordinates. Fig. 8.4b shows the decrease in lobe width and the increase of the number of grating lobes with increasing k; Fig. 8.4c shows the radiation pattern for large k.
In a pattern like that of Fig. 8.4c nulls appear at

$$N\Phi/2 = \pi + 2n\pi \ , \quad n = 1, 2, ... \ . \tag{8.26}$$

Therefore with $N = 2$ and assuming

$$\sin \vartheta \approx \vartheta \tag{8.27}$$

the angle of incidence from (8.20) is

$$\vartheta_0 = (\lambda/D)\{n + (1/2)\} = (n + 1/2)/k \ /\text{rad.} \ . \tag{8.28}$$

The angles of maxima can be obtained from

$$N\Phi/2 = 2m\pi \ , \quad m = 1, 2, 3, ... \tag{8.29}$$

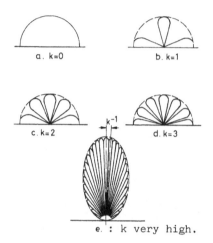

*Fig. 8.4.   Radiation (array) pattern of a
twin interferometer with varying
baseline (inter-element distance).*

and can be found at

$$\vartheta_m = m/k \ /\text{rad.} \ . \tag{8.30}$$

The maxima and minima of the pattern show a mutual distance of $1/k$ rad.. As the lobes are narrow it is legitimate to take the angular distance between nulls as the half-power beam width or "lobe-width" as

$$\vartheta = 1/k = \lambda/D \ . \tag{8.31}$$

By introducing (8.31) in (8.20) and again using the approximation

$$\sin \vartheta \approx \vartheta \tag{8.32}$$

we obtain

$$\Phi_{\max} = 2\pi \ . \tag{8.33}$$

The multilobe pattern of an interferometer causes difficulties, because it is unknown which of the grating lobes sees the target. Although in principle an absolute phase in the range $0 \le \Phi \le 2\pi$ (or $0 \le \Phi \le \pi$ if measured only to one side) is found, it is unknown how often the phase turns through 2π corresponding to the angular width of one lobe. Therefore each measurement has multiple solutions. In order to take care of all possible phase shifts one must add a factor $2n\pi$ to the phase difference evaluated in equation (8.20)

$$\Phi_{Bn} = \Phi_D + 2n\pi \ , \qquad n = 1, 2, \dots \ . \tag{8.34}$$

The measured phase difference Φ_{D_n} corresponds to a possible number of angles of incidence $\vartheta_n \ (\le 1)$

$$\sin \vartheta_n = k^{-1} \left\{ (\Phi_D/2\pi) + n \right\} .$$

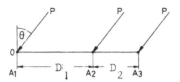

Fig. 8.5. Three-element interferometer.

By adding a further element (see chapter 7.11) as in Fig. 8.5 the ambiguity may be removed if the new baseline D_1 is different from D_2. Analog to the relations (8.20), (8.34) there exists a phase difference to the new antenna of

$$\Phi_{D_1} = (2\pi/\lambda) \sin \vartheta , \tag{8.36}$$

and with the ambiguous Φ_{D_m}

$$\Phi_{D_m} = \Phi_{D_1} + 2m\pi , \qquad m = 1, 2, \dots \tag{8.37}$$

and therefore

$$\sin \vartheta_m = (\lambda/D_1) \left\{ (\Phi_{D1}/2\pi) + m \right\} . \tag{8.38}$$

The new unambiguous value of ϑ is

$$\vartheta = \vartheta_n = \vartheta_m . \tag{8.39}$$

n and m have to be determined by the measurement of Φ_{D_2} (8.34) and Φ_{D_1} (8.36). From (8.39)

$$\sin \vartheta_n = \sin \vartheta_m \tag{8.40}$$

and from (8.35) and (8.38)

$$m = (D_1/D_2)n + (1/2\pi) \left\{ (D_1/D_2)\Phi_{D_2} - \Phi_{D_1} \right\} . \tag{8.41}$$

D_1 and D_2 should be chosen in such a way that potential ambiguities, i.e. inaccurate determinations of n and m, are outside the main lobes of the antenna elements.

The linear array or triplet interferometer consisting of the elements A_1 A_2 A_3 (Fig. 8.5) can be understood as a superposition of the two twin interferometers A_1 A_2 and A_2 A_3. For the A_1 A_2 array the lobe spacing is λ/D (rad) and λ/D_a is the 3-db-beam width (equation 8.24, 8.31), if D_a is the antenna diameter; from this the number of maxima in the antenna main lobe is

$$z = 1 + (D_{1,2}/D_a) . \tag{8.42}$$

If the directivity of the antenna is high, i.e. λ/D_a small, then the number of non-unique maxima in the main lobe is only a fraction of this number, namely

$$z = 1 + (\eta D_{1,2}/D_a) , \qquad \eta < 1 . \tag{8.43}$$

This may be recognized in Fig. 8.4c, which is valid for isotropic radiators (8.31). Interferometer elements, on the other hand, with a finite diameter D_a ($\gg \lambda$) show an increasing multiplicity of beams with increasing frequency, which is expressed by the D_a in the denominator of (8.42).

It may further be concluded that on each side of the central maximum $\eta D/2D_a$ sidelobes exist, which represent ambiguities. All these considerations are valid also for the interferometer formed by $A_2 A_3$.

If the ratio D_1/D_2 in (8.41) is a proper fraction, which can be written as p/q where p and q are relative, whole prime numbers, the q of the lobe of the $A_2 A_3$ interferometer coincides with the p-th lobe of the $A_1 A_2$ interferometer. This is the first coincidence of lobes since that of the central maximum. In order to avoid errors as a consequence of the ambiguities it is necessary that these coincidences, i.e. the correspondence of directions of lobe axes, occur only for sidelobes which are outside the *main beam* of the individual antenna elements. This leads to the following inequalities (17)

$$p > \eta D_1/D_a \tag{8.44}$$

$$q > \eta D_2/D_a \; . \tag{8.45}$$

If Φ_B is the phase difference between A_1 and A_2 and Φ_D between A_1 and A_3 then the random errors of measurement put limits on the magnitude of p and q, if ambiguous solutions of ϑ are to be avoided.

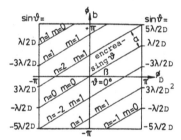

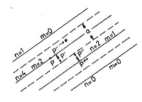

Fig. 8.6.   *Phase relations*
            for a three-
            element inter-
            ferometer (17).

Fig. 8.7.   *To the ambiguity*
            resolution with
            the three-element
            interferometer (17).

Fig. 8.6 shows the phase relations for various ϑ drawn in the $\Phi_{D_1} - \Phi_{D_2}$-plane for $p = 3$ and $q = 5$. The slope of the diagonals is

$$x = p/q = \tan \beta \tag{8.46}$$

and their spacing on the Φ_{D_1} axis

$$\Delta\Phi_{D_1} = 2\pi/q \; . \tag{8.47}$$

The distance between the diagonal is

$$a = \Delta \Phi_D \cos \beta = (2\pi/q) \cos \beta = \frac{2\pi}{q\sqrt{1+x^2}} = \frac{2\pi}{\sqrt{p^2+q^2}} \ . \tag{8.48}$$

The practical value of the $\Phi_{D1} - \Phi_{D2}$-diagram is as follows:
If the measurement of the phase differences Φ_{D1} and Φ_{D2} was made without error, these define a point on one of the diagonals. This then corresponds without ambiguity to a single value of ϑ. If on the other hand the measurement of Φ_{D1} and Φ_{D2} is not error-free, then point P (Φ_{D1}, Φ_{D2}) will be at the location P' which does not lie on a diagonal (Fig. 8.7). A simple method to determine ϑ is then to draw a perpendicular through P' to the nearest diagonal. At the foot of this perpendicular point P^* can be drawn which need not coincide with P. If P' is inside a band of width which symmetrically spans the diagonal containing point P, then P^* lies on the same line as P and the error is small and is a continuous function of the errors of measurement. Magerum (17) calls this "topological uncertainty". If a measurement including errors yield a point P'', the foot of the perpendicular on the diagonal *next* to P'' is P^{**}, which is no longer on the same diagonal as P. In this case the error is considerable and no longer is a continuous function of the errors: an ambiguity exists. For a Gaussian error distribution the *probability* that P' lies inside an unambiguous strip is (32,59,63)

$$\int_{-a/2}^{+a/2} p(x)dx = \Phi\left(\frac{a}{2\sigma}\right) = \Phi\left(\frac{\pi}{\sigma\sqrt{p^2+q^2}}\right) =$$

$$= \frac{1}{\sqrt{2\pi}} \int_{-a/2}^{+a/2} e^{-(1/2)(x/a)^2} dx \ . \tag{8.49}$$

Since

$$\int_{-\infty}^{-\infty} p(x)dx = 1 \tag{8.50}$$

the probability for P' to lie outside the band of width a, i.e. that an ambiguous solution will result, is

$$P_A = 1 - \int_{-a/2}^{+a/2} p(x)dx \ . \tag{8.51}$$

where

$x = X - \bar{X}$	= deviation of a quantity $X(t)$ obeying a *Gaussian probability distribution (82)*
$\sigma^2 = \overline{(X - \bar{X})}^2$	= mean square deviation or *variance* or scattering of errors
σ	= standard deviation.

For a given σ each *probability of ambiguity* corresponds to a value p^2+q^2 and to a boundary condition which states that the probability for the occurrence of an ambiguity must be less than some value A. The relation between the probability of inaccurate resolution of an ambiguity and $p^2 + q^2$ is shown in Fig. 8.8.

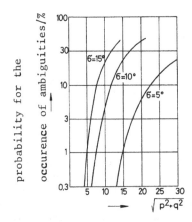

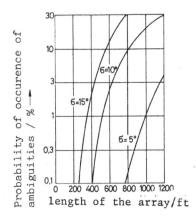

Fig. 8.8.   *Probability of an*   Fig. 8.9.   *Probability of an*
inaccurate ambiguity   *inaccurate ambiguity*
resolution (17).   *resolution (17).*

The total length of an interferometric antenna array is

$$L = D_1 + D_2 \tag{8.52}$$

and with (8.44) and (8.45)

$$D_1 < p\,D_a/\eta \tag{8.53}$$

and

$$D_2 < q\,D_a/\eta \tag{8.54}$$

yields

$$L < (D_a/\eta)(p + q) \ . \tag{8.55}$$

Maximizing the antenna length L for σ means also maximizing $p + q$ for a fixed value of $p^2 + q^2$. This implies $p = q$, which is inadmissible since the system is ambiguous for $D_1 = D_2$. It is, however, possible to use values for p and q close to one another, for instance $q = p + 1$. This leads to a maximum length of the array of

$$L < (D_a/\eta)(2p + 1) \ . \tag{8.56}$$

If the directional error is $\pm 1/4$ of the half-power beamwidth at $\eta = 1/2$ and if the antennas are parabolic reflectors of $D_a = 25.$ ft the maximum array length is $L = 50 + 100\,p$ (feet). For a given value of p the probability of inaccurate ambiguity resolution can be found from Fig. 8.8. Fig. 8.9 shows the probability for the occurrence of ambiguities for total length L (equation 8.52) and $D_a = 25.$ ft. After resolving these ambiguities the exact value of ϑ may be obtained from the phase difference on the longest baseline (17)

$$\sin \vartheta = (\lambda_0/2\pi)\,(\Phi_{D^1} + \Phi_{B^2})/(D_1 + D_2) \tag{8.57}$$

or

$$\sin \vartheta = \{m + n + (\lambda_0/2)\}\,(\Phi_{D^1} + \Phi_{D^2})/(D_1 + D_2) \ . \tag{8.58}$$

The attendant error is (17)

$$\sigma_\vartheta = (\lambda_0/2) \left(\sqrt{2}\,\sigma_\Phi / \{ (D_1 + D_2) \cos \vartheta \} \right), \tag{8.59}$$

where σ_Φ is the error of the phase measurement. Fig. 8.10 shows equation (8.59) in graphic form for a centre frequency in X-band and $D_a = 25.$ ft.

Three chief sources of phase error σ_Φ exist (17):

1. atmospheric variations

2. errors in the measuring system

3. errors caused by misalignment of the phase centres of the array elements.

The combination of these error sources can lead to ambiguities.

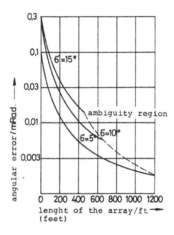

Fig. 8.10. *Measuring accuracy of the system (17).*

8.2.4. Resolving Power and Ambiguity Function

8.2.4.1. Radar Receiver Optimization

The model of a radar receiver used in the following discussion is based on a linear filter followed by a detector. The detector compares the envelope of the output voltage of the filter, the function $s_0(t)$, with a threshold voltage. If this arbitrarily chosen threshold voltage is exceeded the receiver reports the existence of an echo signal. The level of the threshold influences the *probability* of both correct and *false echo detections*. The filter has the task of ensuring the greatest possible signal-to-noise power ratio. Since in principle a detection can be obtained when the signal exceeds the threshold for an arbitrarily short time. The signal-to-noise ratio does not need to be at a maximum during the whole signal duration, but only at some particular time. The filter must maximize the *signal-to-noise ratio*

$$R_f = s_0^2(t)_{\mathrm{max}}/N', \tag{8.60}$$

where $|s_0(t)_{max}|$ corresponds to the maximum value of the signal voltage at the output and N' to the mean value of the noise power at the receiver output (209). The solution to this maximizing problem is given by the *matched filter* (77), which represents the optimal method of distinguishing signals in the presence of Gaussian noise. The *transfer function* in the frequency domain can be written as

$$H(f) = V_a S^*(f) \exp(-2j\pi f t_1) \tag{8.61}$$

with

$$S(f) = \int\limits_{-\infty}^{+\infty} s(t) \exp(-2j\pi f t)dt \tag{8.62}$$

as a spectrum of an input time signal $s(t)$

$S^*(f)$ the complex conjugate function of $S(f)$,
t_1 the signal duration,
V_a the maximum "amplification" of the filter
            (in general equal to one).

Except for the phase shifting factor $\exp(-j2\pi f t_1)$ which changes continuously with the frequency and produces a continuous time shift, the function $H(f)$ corresponds to the complex conjugate spectrum of the input signal. The reason for this delay is, that no signal can appear at the output of the filter until the complete input signal has been received.

If the frequency spectra $S(f)$ and $H(f)$ are divided into amplitude spectra $|S(f)|$, and $|H(f)|$ and phase spectra $\exp\{-j\varphi_s(f)\}$, $\exp\{-j\Phi_{of}(f)\}$, omitting the constant V_a (equation 8.61) yields

$$|H(f)|.\exp\{-j\Phi_{of}(f)\} = |S(f)| \exp\{-j[\Phi_s(f) - 2\pi f t_1]\} . \tag{8.63}$$

Hence it follows that

$$|H(f)| = |S(f)| \tag{8.64}$$

$$\Phi_{of}(f) = -\Phi_s(f) + 2\pi f t_1 , \tag{8.65}$$

that the amplitude spectra are equivalent to each other and the phase spectrum of the matched filter is equal to the negative of the signal phase spectrum plus a phase shift proportional to the frequency.

The matched filter can also be characterized by the impulse response (54,55,56,58,60,61, 77,68):

The *response* to a *single pulse* is:

$$s_0(t) = \int\limits_{-\infty}^{+\infty} S_0(f) \exp(j2\pi f t)df . \tag{8.66}$$

With

$$S_0(f) = H(f)S(f) \tag{8.67}$$

and

$$s(t) \triangleq \delta(t) \quad \circ\!\!-\!\!\circ \quad 1 \triangleq S(f) , \tag{8.68}$$

(where o−o is the *transformation symbol* between time domain and frequency domain), from which follows

$$s_0(t) \triangleq h(t) = \int_{-\infty}^{+\infty} H(f) \exp(j2\pi ft) df \qquad (8.69)$$

with

$$H(f) = S(f) . \qquad (8.70)$$

By introducing equation (8.61) in equation (8.69) we obtain

$$h(t) \triangleq V_a \int_{-\infty}^{+\infty} S^*(f) \exp\left\{-j2\pi f(t_1 - t)\right\} df . \qquad (8.71)$$

Since

$$S^*(f) = S(-f) \qquad (8.72)$$

equation (8.71) changes to

$$h(t) = V_a \int_{-\infty}^{+\infty} S(-f) \exp\left\{j2\pi(-f)(t_1 - t)(-1)\right\}(-df) . \qquad (8.73)$$

With the substitution

$$-f = f_z \qquad (8.74)$$

finally

$$h(t) = V_a \int_{-\infty}^{+\infty} S(f_z) \exp\left\{j2\pi f_z(t_1 - t)\right\} df_z \qquad (8.75)$$

and

$$h(t) = V_a s(t_1 - t) . \qquad (8.76)$$

The impulse factor shows also that the output function of the filter is equal to the reflected image of the input functions in the time domain (Fig. 8.11).

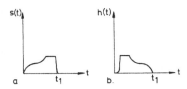

Fig. 8.11.   To the matched filter
a.   recept signal form $s(t)$
b.   pulse response $h(t)$ of the matched
filter on the input signal $s(t)$.

Now the output signal of the filter matched to the expected signal $s_{Ref}(t)$ must be calculated by convolution (72,73,68) of the *echo signal function* $s(t)$ with the impulse response (8.71) using (8.67). Apart from a time delay $s_{Ref}(t)$ is merely a copy of the transmitted signal which is stored in the receiver.

$$S(f)H(f) \quad \circ\!\!-\!\!\circ \quad s(t) * h(t) \tag{8.77}$$

$$s_0(t) = s(t) * h(t) = \int\limits_{-\infty}^{+\infty} s(\tau)h(t-\tau)d\tau . \tag{8.78}$$

using

$$h(t-\tau) = V_a s_{Ref.}(t_1 - t + \tau) \tag{8.79}$$

it follows that

$$s_0(t) = V_a \int\limits_{-\infty}^{+\infty} s(\tau)s_{Ref.}(t_1 - t + \tau)d\tau . \tag{8.80}$$

If we disregard causality and set the arbitrary constant t_1 in (8.80) to zero, then

$$s_0(t) = V_a \int\limits_{-\infty}^{+\infty} s(\tau)s_{Ref.}(\tau - t)d\tau , \tag{8.81}$$

and it can be seen that the output signal of the matched filter is equal to the *crosscorrelation* function (ccf) of the echo signal and the reference signal. Fig. 8.14 shows the output function of a matched filter for a rectangular impulse (Fig. 8.13) at the input. As the reference signal corresponds exactly to the echo signal delayed by the transmission time the output signal of the matched filter is also equal to the *auto correlation* function (acf) of the signal.

The output voltage of a filter with the transfer function $H(f)$ is given by the equation (8.66) and (8.67)

$$|s_0(t)| = \left| \int\limits_{-\infty}^{+\infty} S(f)H(f)\exp(j2\pi ft)df \right| . \tag{8.82}$$

Following (58,77) the mean value of the output noise power is

$$N' = \frac{N_0}{2} \int\limits_{-\infty}^{+\infty} |H(f)|^2 df , \tag{8.83}$$

where N_0 is equal to the input noise power in W/Hz. The factor $1/2$ is due to the fact that the limits of the integral are $+\infty$ and $-\infty$ while N_0 is only defined in the *positive* frequency domain.

If these two equations are introduced into equation (8.60) with the assumption that the value $|s_0(t)|^2$ has a maximum at the instant t_1, the signal-to-noise ratio then becomes

$$R_f = \frac{\left| \int\limits_{-\infty}^{+\infty} S(f)H(f)\exp(j2\pi ft_1)df \right|^2}{\dfrac{N_0}{2} \int\limits_{-\infty}^{+\infty} |H(f)|^2 df} . \tag{8.84}$$

The equation (8.84) may be formulated in a more understandable way by use of the Schwarz' inequality (61,72,75,77,82):

The *Schwarz' inequality* for the case of two complex functions $\underline{A}$ and $\underline{B}$ is

$$\int \underline{A}^* \underline{A} \, dx \int \underline{B}^* \underline{A} \, dx \geq \left| \int \underline{A}^* \underline{B} \, dx \right|^2 .$$

(8.85)

Equality exists in the above equation if

$$\underline{A} = k.\underline{B}$$

(8.86)

is fulfilled, with k a constant. Supposing further that

$$\underline{A}^* = S(f) \exp(2\pi f t_1)$$

(8.87)

and

$$\underline{B} = H(f) ,$$

(8.88)

since

$$\int \underline{A} \, \underline{A}^* \, dx = \int |\underline{A}|^2 dx$$

(8.89)

then

$$R_f \; \lesseqgtr \; \frac{\displaystyle\int_{-\infty}^{+\infty} |H(f)|^2 df \int_{-\infty}^{+\infty} |S(f)|^2 df}{\displaystyle\frac{N_0}{2} \int_{-\infty}^{+\infty} |H(f)|^2 df}$$

(8.90)

$$R_f \; \lesseqgtr \; \frac{\displaystyle\int_{-\infty}^{+\infty} |S(f)|^2 df}{\displaystyle\frac{N_0}{2}} \; .$$

(8.91)

By application of *Parceval's theorem* (60,72,73,75,77,82) the following result may be obtained

$$\int_{-\infty}^{+\infty} |S(f)|^2 df = \int_{-\infty}^{+\infty} |s(t)|^2 dt .$$

(8.92)

Here

$$\int_{-\infty}^{+\infty} |s(t)|^2 dt = E' ,$$

(8.93)

which is equal to the signal energy. We can rewrite equation (8.91) as

$$R_f \; \lesseqgtr \; 2E'/N_0 .$$

(8.94)

The maximization we require is achieved if

$$R_f = 2E'/N_0 , \tag{8.95}$$

which means that equation (8.86) is fulfilled. Thus

$$S^*(f).\exp(-j2\pi f t_i) = k.H(f) . \tag{8.96}$$

This is identical with (8.61) if

$$k = 1/V_a . \tag{8.97}$$

Equation (8.95) shows that it is characteristic of a matched filter, that the maximum signal-to-noise ratio is independent of the form of the input signal. The problem of maximizing the target detection probability of the system for a given probability of false detection can be solved by a suitable choice of the filter.

Additionally, the resolution of the system can be optimized by the choice of signal form without compromising the optimal detection probability. So far we have assumed that the angular resolution is equal to the half-power beamwidth of the antenna main lobe which for an interferometer is equal to the central grating lobe of the beam. We must now consider a more *general* definition of *resolution* as the ability clearly to resolve the echo from a target in the presence of other targets, even when echos of equal strength from these targets overlap. The probability that a *matched receiver* falsely interprets two overlapping echos as a single echo, or reports two targets when only one is present is for the case of large signal-to-noise ratio (75,209,213,214,215,219)

$$Q \approx \exp\left(\left(\frac{1}{16} R_f 1 - \chi(\Delta\tau, f_d)\right)^2\right) . \tag{8.98}$$

Here $\chi(\Delta\tau, f_d)$ is the *ambiguity function* introduced by Woodward (75), which is discussed in detail in the following section. $\Delta\tau$ corresponds to the time difference and Δf_d to the frequency between two targets. If both the frequency and time difference between the targets approach zero, the probability of confusion approaches unity, since the normalized ambiguity function is unity for this case. Equation (8.98) can be considered as an implicit definition of the resolution, since it expresses the minimum resolvable element $\Delta\tau$, Δfd as a function of signal-to-noise ratio R_f and signal form expressed by χ for a given probability Q. For the purpose of optimization it is only necessary to consider the signal waveform.

8.2.4.2. Derivation of the Ambiguity Function (af)

The ambiguity function must first be determined starting from the autocorrelation function (acf) and the reference signal stored in the receiver, which is a copy of the transmitted signal delayed by the propagation time τ. The acf is

$$r(\tau, y) = \int\limits_{-\infty}^{+\infty} s(t)s(t - \tau)dt . \tag{8.99}$$

To simplify the following derivation the Doppler effect is initially neglected and signal functions are introduced by regarding the real signal to be processed as the real part of a hypothetical, complex signal $\psi(t)$. The signal $\psi(t)$ originates from the suppression of the components of the amplitude spectrum $s(f)$ corresponding to $s(t)$ for $f < 0$ and by doubling the amplitudes at $f > 0$ (Fig. 8.12).

This yields

$$s(t) = \text{Re}\,\{\Psi(t)\}\ . \tag{8.100}$$

$\psi(t)$ can be separated into the modulation function

$$u(t) = a(t)\exp\{j\varphi(t)\} \tag{8.101}$$

and the *carrier oscillation* $\exp(j\omega_0 t)$

$$s(t) = \text{Re}\,\{a(t)\}\exp\{j\varphi(t)\}\exp(j\omega_0 t) =$$
$$= \text{Re}\,\{u(t)\exp(j\omega_0 t)\}\ . \tag{8.102}$$

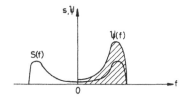

Fig. 8.12.   To the definition of an
"analytical signal" $\Psi(f)$, (61).

If the carrier frequency $f_0 = \omega_0/2\pi$ in (8.102) is higher than half the bandwidth (75)

$$f_0 > \Delta f/2 \tag{8.103}$$

then we write

$$\Psi(f) = \begin{cases} 2S(f), & f > 0 \\ 0, & f < 0. \end{cases} \tag{8.104}$$

After application of the *convolution theorem* (67,68,73) for the two complex signal functions

$$\psi_1(t)\psi_2(t)\ \circ\!-\!\circ\ \Psi_1(f) * \Psi_2(f) \tag{8.105}$$

and assuming (72,73)

$$\psi^*(t)\ \circ\!-\!\circ\ \Psi(-f) \tag{8.106}$$

we obtain, writing the convolutions in full

$$\int_{-\infty}^{+\infty} \Psi_1(f')\Psi_2^*(-f+f')df' = \int_{-\infty}^{+\infty} \psi_1(t)\,\psi_2^*(t)\exp(j2\pi ft)dt \tag{8.107}$$

and at $f = 0$

$$\int_{-\infty}^{+\infty} \psi_1(t)\,\psi_2^*(t)dt = \int_{-\infty}^{+\infty} \Psi_1(f')\,\Psi_2^*(f')df\ . \tag{8.108}$$

Equation (8.108) corresponds to Parseval's theorem (63).
        Similarly, considering two real signal functions $s_1(t)$ and $s_2(t)$

$$\int\limits_{-\infty}^{+\infty} s_1(t)\, s_2(t) dt = \int\limits_{-\infty}^{+\infty} S_1(f)\, S_2^*(f) df \qquad (8.109)$$

or

$$\int\limits_{-\infty}^{+\infty} S_1(f)\, S_2^*(f) df = \int\limits_{-\infty}^{0} S_1(f)\, S_2^*(f) df + \int\limits_{0}^{+\infty} S_1(f)\, S_2^*(f) df =$$

$$= -\int\limits_{0}^{-\infty} S_1(f)\, S_2^*(f) df + \int\limits_{0}^{+\infty} S_1(f)\, S_2^*(f) df \; . \qquad (8.110)$$

With the substitution $f = -f$ in the first term on the right-hand side of equation (8.110) and remembering (8.72)

$$S^*(f) = S(-f) \qquad (8.111)$$

we obtain

$$\int\limits_{0}^{+\infty} S_1(f)\, S_2^*(f) df = \int\limits_{0}^{+\infty} S_1^*(-f)\, S_2^*(-f) df + \int\limits_{0}^{+\infty} S_1(f)\, S_2^*(f) df =$$

$$= \int\limits_{0}^{+\infty} \{ S_1^*(f)\, S_2(f) + S_1(f)\, S_2^*(f) \}\, df =$$

$$= 2\,\mathrm{Re}\left\{ \int\limits_{0}^{\infty} S_1(f)\, S_2^*(f) df \right\} ; \qquad (8.112)$$

but from equation (8.104) it follows that

$$\int\limits_{-\infty}^{+\infty} S_1(f)\, S_2^*(f) df = 2\,\mathrm{Re}\left\{ 1/4 \int\limits_{-\infty}^{+\infty} \Psi_1(f)\, \Psi_2^*(f) df \right\} \qquad (8.113)$$

and with (8.108) and (8.109) we obtain

$$\int\limits_{-\infty}^{+\infty} s_1(t)\, s_2(t) dt = 1/2\,\mathrm{Re}\left\{ \Psi_1(f)\, \Psi_2^*(f) df \right\} =$$

$$= 1/2\,\mathrm{Re}\left\{ \int\limits_{-\infty}^{+\infty} \psi_1(t)\, \psi_2^*(t) dt \right\} . \qquad (8.114)$$

By applying the theorem of equation (8.114) to equation (8.99)

$$r(\tau, f_d) = \int\limits_{-\infty}^{+\infty} s(t)\, s(t-\tau) dt = 1/2\,\mathrm{Re}\left\{ \int\limits_{-\infty}^{+\infty} \psi(t)\, \psi^*(t-\tau) dt \right\} . \qquad (8.115)$$

Introducing the complex echo signal frequency shifted by the Doppler effect

$$\psi(t) = u(t)\,e^{j2\pi(f_0+f_d)t} \tag{8.116}$$

and the stored and time-delayed copy of the transmitted signal

$$r(t-\tau) = u(t-\tau)\,e^{j2\pi(f_0+f_d)t} \tag{8.117}$$

in equation (8.115) yields

$$r(\tau, f_d) = 1/2\,\mathrm{Re}\left\{\int_{-\infty}^{+\infty} u(t)\,e^{j2\pi f_0 t}\,e^{j2\pi f_d t}\,u^*(t-\tau)\,e^{-j2\pi f_0 t}\,e^{j2\pi f_0\tau}\,dt\right\} =$$

$$= 1/2\,\mathrm{Re}\left\{e^{j2\pi f_0\tau}\int_{-\infty}^{+\infty} u(t)\,u^*(t-\tau)\,e^{j2\pi f_d t}dt\right\} =$$

$$= 1/2\,\mathrm{Re}\left\{e^{j2\pi f_0\tau}\,\chi(\tau, f_d)\right\} \tag{8.118}$$

which is Woodward's ambiguity function (a.f.) (75).

$$r(\tau, f_d) = \int_{-\infty}^{+\infty} u(t)\,u^*(t-\tau)\,e^{j2\pi f_d t}\,dt\ . \tag{8.119}$$

Originally Woodward derived his a.f. from the mean square deviation between the signals $s(t)$ and $s(t-\tau)$, i.e. using (75).

$$\varepsilon^2 = \int_{-\infty}^{+\infty} |\psi(t) - \psi(t-\tau)|^2 dt\ =$$

$$= \int_{-\infty}^{+\infty} \{|\psi(t)|^2 + |\psi(t-\tau)|^2\}\,dt - 2\,\mathrm{Re}\left\{\int_{-\infty}^{+\infty} \psi(t)\,\psi^*(t-\tau)dt\right\}\ . \tag{8.120}$$

Here, the first term stands for the energy contained in both signals and with help of the second term the a.f. can be derived analogous to equation (8.115).

In section 8.2.4.1. the output signal of a matched filter was interpreted as the crosscorrelation of an echo signal and a reference signal or as autocorrelation when one considers the similarity of the signals. From comparison of equations (8.81) and (8.99) or from the expressions derived from equation (8.118) it may be concluded that the a.f. is equal to the complex, time-inverted envelope of the matched filter output signal as a function of delay and Doppler shift, i.e. of distance and velocity of the target (209,213,214). The real output function is found by considering the carrier and forming the absolute value of $\chi(\tau, f_d)\exp(j\pi f_0\tau)$. The real envelope is then $|\chi(\tau, f_d)|$ (75).

The absolute value of $\chi(\tau, f_d)$ represents a surface stretched over the frequency-time plane. Every cut through the parallel to the time axis describes an output signal, whose Doppler shift compared to the reference signal just corresponds to distance from the time axis. We can consider a bank of matched filters, where each filter is tuned to a certain Doppler shift, so that the matched filter receiver is able to react to an unknown Doppler shift. Then the a.f. describes the whole two-dimensional output function of such a receiver as a function of delay τ and Doppler shift f_d. When $|\chi|$ is distributed over the $\tau - f_d$-plane, $|\chi|$ represents the *combined* output function of *all* filters of a Doppler bank with infinitesimally small f_d-distance between the individual f_d values. Normally $|\chi|$ is drawn in normalized form, such that $|\chi(0,0)| = 1$. At a target velocity of zero the function

$$\chi(\tau,0) = \int\limits_{-\infty}^{+\infty} u(t)\,u^*(t-\tau)dt \tag{8.121}$$

corresponds to the output function of the filter which is tuned to zero and this signal is identical with the acf. This acf is triangular (Fig. 8.13) for a rectangular impulse and corresponds in demodulated form (209,213,214) to the function shown in Fig. 8.14 the half-power width (h.p.w.) and therefore the smallest *resolvable distance difference* is equal to the impulse duration (16,68). This provides us with another definition of resolution.

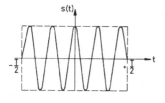

Fig. 8.13.   Sine-formed carrier
with rectangular
envelope (61).

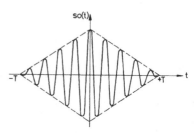

Fig. 8.14.   Output function of
a matched filter
for a rectangular-
formed input
pulse (61).

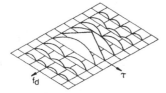

Fig. 8.15.   Ambiguity function
for different
values: $f_d =$
const. (75,77).

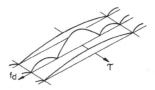

Fig. 8.16.   Ambiguity function
for different
values: $\tau =$
const. (75,77).

One can plot the output functions of all the individual matched filters on one diagram, as a function of τ at fixed values of f_d in Fig. 8.15 and as a function f_d in Fig. 8.16. The plot for $\tau = 0$ is the amplitude spectrum of a rectangular pulse. Fig. 8.17 combines the two previous plots to show the ambiguity function for a rectangular pulse.

A three-dimensional representation of the function is shown in Fig. 8.19. Each target which is in the radar beam is assigned to an ambiguity function, the amplitude and phase of the echo signal being dependent on the amplitude and phase of the reflection coeffi-cient of the target. The single ambiguity functions may then be superimposed to obtain the required receiver output function for all targets. A cut through the ambiguity surface

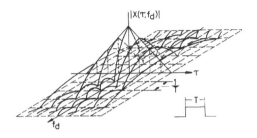

Fig. 8.17.   Ambiguity function of a rectangular
pulse as a superposition of
Figs. 8.15 and 8.16 (216,217).

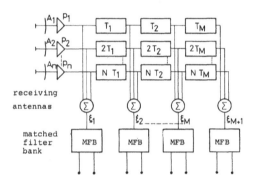

Fig. 8.18.   Linear antenna array with
delay networks (218).

at a particular frequency shift f_d shows the output function of the filter tuned to this
frequency which is required in order to detect the targets present. Inspection of the
ambiguity function shows which peaks are central lobes of the output function, i.e. which
can be regarded as resolvable targets. The *probability of confusion* is given by equation
(8.98). The required form of the ambiguity surface depends on the application of the radar
system. If echoes are to be resolved both in distance and velocity, then the ambiguity
surface has the shape of a *"needle"* or *"thumbtack* ambiguity" (Fig. 8.20a) (75). If the
Doppler resolution is not of interest, then the shape of a wedge of finite curvature (Fig.
8.20b) is sufficient. In any case it is important to minimize secondary maxima as far as
possible, because otherwise the measurement is ambiguous; a high secondary maximum
at a position $\tau = t$ and $f_d = f_{d_1}$ causes by equation (8.98) a high *confusion probability*
Q (82) at distance $\Delta \tau = t_1$ and $d_d = f_d$ from the origin. The a.f. is always fixed for
a particular signal form; therefore $\chi(\tau, f_d)$ is defined as: "ambiguity function of a signal
form". Unfortunately, it is not possible to find by calculation the signal form from the

given ambiguity surface (75), only the inverse procedure is possible, a process of successive approximation leads to a signal form corresponding to a desirable ambiguity function.

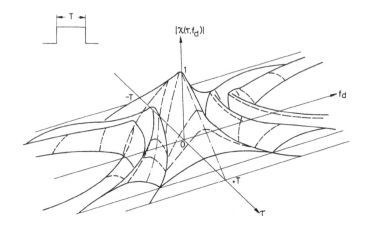

Fig. 8.19.   Ambiguity function of a rectangular pulse of duration T (218).

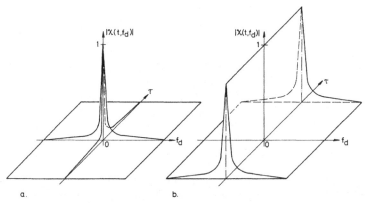

Fig. 8.20.   Ideal form of the ambiguity surface for
a. high time and frequency resolution
b. only time resolution.

8.2.4.3. Pulse Compression

If a system uses signals formed by modulating a carrier, then generally the "sharpness" (width) of the matched filter output signal is inversely proportional to the effective signal band width (82, p. 117–120) (75). Consequently, if the matched filter output signal becomes a sharp "tack", then the signal bandwidth must be broad.

   This requirement of a broad signal bandwidth puts constraints on the radar system, for example if pulse duration is decreased to increase the signal bandwidth. As the signal-to-noise ratio of a target signal and the measuring accuracy depend on the energy content of the signal used, the transmitted power must be increased proportionally to the shrinking *pulse duration* (see also: (8.207)) if the transmitted energy is to remain constant. However, the *peak power* of the transmitted energy sets limits to the pulse duration. However, a useful Doppler resolution is related to a long pulse duration.

In order to obtain both *distance resolution* and adequate pulse length, the pulses are frequency- or phase-modulated or also frequency-switched or phase-switched (209,215). In the case of frequency modulation of a pulse of amplitude A and duration T the transmitting frequency is varied linearly between f_1 and f_2 with $f_2 > f_1$ (Fig. 8.21a).

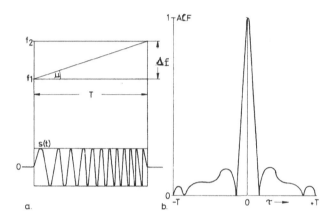

Fig. 8.21.   Pulse compression by linear frequency
modulation (209).

After reflection the echo signal is applied to a *pulse compression filter* using surface acoustic wave (SAW (86)) devices, in which the transmission velocity of the signal is a function of the frequency, so that the higher frequency parts of the pulse which arrive later are accelerated (78). Thus the energy contained in the original pulse of length T is now in a shorter one of a duration of about $1/\Delta f$ where $\Delta f = f_2 - f_1$ (8.122) is the frequency sweep. The original peak power of the signal is increased after passing through the pulse compression filter of the receiver by a factor $\Delta f.T$. The corresponding form of the a.c.f. is shown in Fig. 8.21b. The time-bandwidth product $\Delta f.T$ is also called the *pulse compression ratio* (75).

The most important difference between a swept-frequency *signal matched* to a pulse-comparison filter and a simple pulse is that the first has a time-bandwidth product much greater than one. This technique can be extended to use transmission and reception of nonlinearly frequency-modulated pulses.

Another technique for pulse compression is *binary-switched RF phase*. Here a long pulse of duration T is divided into $\Delta_0 f.T$ equally long intervals of duration $1/\Delta f$, where the phase $0°$ or $180°$ is assigned to each interval. The modulation function can be completely described by a sequence of binary elements of the form $+$, $-$, $-$, etc., which represents the phase. Such a sequence is called: a *code* (58,56,57,80,81). The compression ratio is equal to the number of elements of this code; a.c.f. also depends on the code. The code should be chosen to give as great a ratio of main maximum to secondary maximum in

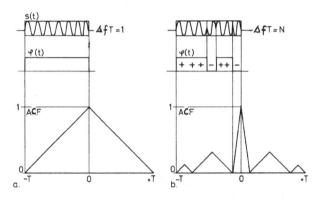

Fig. 8.22.   *Modification of the autocorrelation function by phase modulation (209).*

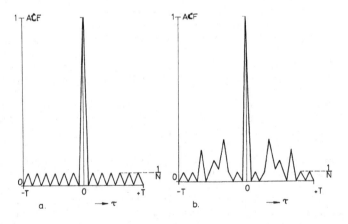

Fig. 8.23.   *Autocorrelation function of a Barker code $N = 13$ and pseudo-noise code $N = 15$ (209).*

order to avoid false detections. For a rectangular pulse without frequency modulation the a.c.f. is triangular, as Fig. 8.22a, and the half power width is equal to the pulse duration $T = 1/\Delta f$. For phase- or frequency-modulated pulses the a.c.f. has several secondary maxima (Figs. 8.21b, 8.22b).

Here also, the smallest resolvable distance corresponds to about the half power width of the main maximum (75) but this is only a fraction of the half power width of the unmodulated signal. Very useful correlation functions result from the Barker codes (56,209) for which the ratio between main and secondary levels is equal to the number of elements. Barker codes can contain up to 13 elements (1986), see Fig. 8.23a.

In order to achieve higher compression ratios, codes of higher element numbers must be used.

For practical applications with large numbers of elements, *pseudo-random codes* (75,76,220) are very useful. These codes simulate white noise, Fig. 8.23b.

8.2.4.4. Signal Form and Antenna Pattern

First the output function of a linear array of isotropic antenna elements of infinitesimal separation (Fig. 8.24a) is given by (75).

$$\psi(t, \vartheta) = \int_{-\infty}^{+\infty} \psi(t - \tau) I(x) e^{j\Phi(x)} dx \ . \tag{8.123}$$

In equation (8.123) we have substituted $x = l/\lambda_0$ and

$$\psi(t - \tau) = u(t - \tau) e^{j2\pi f_0(t-\tau)} \ , \tag{8.124}$$

is a narrowband signal received by a particular element of this array at position l with

$$\tau = (l/\lambda_0 f_0) \sin \vartheta \ , \tag{8.125}$$

the delay delay due to the distance l of this antenna from the reference element with $\tau = 0$: $I(x) \exp\big(j\Phi(x)\big)$ is the complex *illumination* function (6,15,23,26,159), where the phase term $\Phi(x)$ allows us to attribute a different phase to each array element. The antenna pattern of the individual elements is of no interest and is therefore taken as 1, i.e. isotropic elements are used. After manipulation of equation (8.123) we obtain

$$\psi(t, \tau) = e^{j2\pi f_0 t} \int_{-\infty}^{+\infty} u(t - (x/f_0) \sin \vartheta) \cdot$$

$$\cdot e^{-j2\pi x \sin \vartheta} I(x) e^{-j\Phi(x)} dx \ . \tag{8.126}$$

In the case of a continuous sinewave signal the envelope $u(t)$ is constant and therefore need not be further considered. Then the output function of the antenna array

$$\psi(\vartheta) = \int_{-\infty}^{+\infty} I(x) e^{j\Phi(x)} e^{-j2\pi x \sin \vartheta} dx \ . \tag{8.127}$$

As derived in chapter 5 (equation 5.10) a comparison of equation (8.127) with the Fourier representation of a complex signal (8.68)

$$\psi(t) = \int\limits_{-\infty}^{+\infty} \Psi(f)\, e^{j2\pi f t}\, df \qquad\qquad (8.128)$$

shows that the far-field output function of an antenna is related to the illumination function as the temporal form of a signal is to its frequency spectrum. An insignificant difference between the two cases is that $\sin\vartheta$ instead of ϑ is used in an analogy to time. This is of no importance in practice as the significant lobes are so narrow that the approximation $\sin\vartheta \approx \vartheta$ is valid.

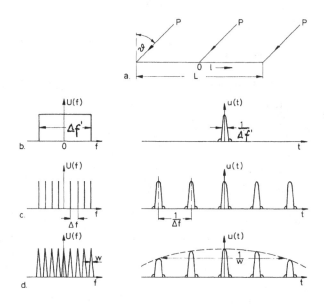

Fig. 8.24.   To the derivation of the radiation pattern
              of a continuous array, a. geometry b., c.,
              d. complex modulation function in the fre-
              quency and time domain (75).

An analogy can also be seen between antenna pattern and signal form. To avoid ambiguities the illumination pattern must be continuous and without discontinuities. Sudden maxima in the illumination pattern $I(x)$ cause high sidelobes just as discontinuities in the amplitude spectrum lead to secondary maxima in the signal form. The sidelobes may be as large as the main lobe, as is the case for an interferometer with element spacing $D > \lambda_0$. These maxima are called "grating lobes" (see section 7.7).

As the preceding section dealt with the matched filter output function in detail, it is now opportune to compare equation (8.127) with one of the functions derived above. First we modify equation (8.98).

For $f_d = 0$, if

$$r(\tau,0) = \mathrm{Re}\left\{\rho(\tau,0)\right\} \qquad\qquad (8.129)$$

$$\rho(\tau,0) = 1/2\, e^{j2\pi f_0} \int\limits_{-\infty}^{+\infty} u(t)\, u^*(t-\tau)dt \;. \tag{8.130}$$

Using the convolution theorem (72,73) we obtain

$$u^*(t-\tau) \; \circ\!-\!\circ \; U^*(f)\, e^{j2\pi f} \tag{8.131}$$

and thus

$$\rho(\tau,0) = 1/2\, e^{j2\pi f_0 \tau} \int\limits_{-\infty}^{+\infty} |U(f)|^2\, e^{j2\pi f\tau}\, df \;. \tag{8.132}$$

The envelope of the real acf of the matched filter output function is then

$$\rho(\tau,0) = 1/2 \left| \int\limits_{-\infty}^{+\infty} u(t)\, u^*(t-\tau)dt \right| = 1/2 \left| \int\limits_{-\infty}^{\infty} |U(f)|^2\, e^{j2\pi f\tau}\, df \right| \;. \tag{8.133}$$

In a comparison of equations (8.127) and (8.133) $I(x)$ takes the place of the energy density spectrum $|U(f)|^2$ of the complex signal envelope. Now we must consider if it is most appropriate to compare equation (8.127) with (8.133) or with (8.128). A radar system is unable to transmit a continuous spectrum down to $f = 0$, which would be a necessity to avoid ambiguous range measurements. Instead a suitable spectrum is superposed on a carrier before the signal is transmitted and so the range resolution depends alone on the *complex* envelope. For an array system similar difficulties can be surmounted by keeping the illumination function continuous along the whole length of the aperture. For this reason $I(x)\exp\big(j\Phi(x)\big)$ corresponds rather to $|U(f)|^2$ than $|\psi(f)|^2$. If $U(f)$ is a periodic spectrum consisting of spectral lines of equal amplitude with spacing Δf then using Fourier analysis we find that $u(t)$ has a series of equal lobes of separation $1/\Delta f$ (Fig. 8.24c), where the single lobes have the same form as the lobe of a continuous spectrum with the same envelope as $U(f)$, (Fig. 8.24b). Therefore regular gaps in the spectrum produce ambiguous secondary maxima in the time domain and hence ambiguities in the range, i.e. transmission time measurements. The one-dimensional array is according to this a linear array of isotropic elements with a *multi-lobed* array pattern, e.g. a grating interferometer. If every spectral line in Fig. 8.24c is enlarged to a certain finite band this results in a configuration which is shown in Fig. 8.24d in the time domain which is formed by multiplication of the signal group of Fig. 8.24c with an envelope whose width corresponds to the reciprocal of the width of the spectral "line". The outer sidelobes are attenuated and may even vanish entirely. Therefore the broader the spectral "line" becomes the smaller is the number of repeated signals. Comparing this behaviour with an extended antenna array, it can be stated that by physically extending its elements, e.g. by enlarging the diameter of a parabolic reflector element the number of ambiguous grating lobes can be decreased.

If gaps in the spectrum are avoided, there exists a single "lobe" in the time domain just as a continuous illumination function produces a narrow main lobe in the radiation pattern of an antenna. The larger the extension of an array or interferometer element, the higher is its directivity and the sharper its main lobe, i.e. the smaller the half-power beam width.

If very high accuracy in angular resolution is specified, the dimensions of the element antennas become so large that practical considerations force us either to abandon continuous illumination of the elements or to make the element separation so small that the occurrence of high secondary lobes is avoided. It is, however, preferable to choose an element distance or baseline of many wavelengths, and to solve the problem of possible ambiguities as for the twin interferometer (see chapters 7.11 and 8.2.3)—by addition of a third element. For even higher element numbers grating lobes and attendant danger of ambiguity can be reduced by choice of an appropriate signal bandwidth. When considering the antenna output function (equation (8.127)) a continuous sinewave signal was assumed so that $u(t)$ could be taken outside the integral as a constant. This is also allowed, if the variations of $u(t)$ during the integration interval are negligible.

This corresponds to the maximum possible time delay when the signal arrives at an antenna array (8.24a)

$$\tau_{\text{max}} = (L/c_0) \sin \vartheta .\tag{8.134}$$

$u(t)$ can only be regarded as constant and therefore having negligible influence on the array characteristic if the signal duration T is considerably longer than τ_{max}. The condition for no influence of signal form on the array characteristic can be formulated for a fixed time-bandwidth product $\Delta f T = 1$ as $\tau \ll 1/\Delta f$

$$\Delta f \ll c/(L \sin \vartheta_0) .\tag{8.135}$$

Here ϑ_0 defines the main lobe width to first null. For $\vartheta = 0$ equation (8.135) is for all cases fulfilled.

As the half-power beam width (HPBW) of the main lobe is given by the relation $\vartheta_m = \lambda_0/L$ the angle corresponding to a full lobe width away from the zenith, i.e. from the axis of the main lobe is $\vartheta_0 = \lambda_0/L$. In this case, if also $\sin \vartheta \approx \vartheta$, then (8.135) becomes

$$\Delta f \ll c/(L\vartheta)\tag{8.136}$$

or

$$\Delta f/f_0 \ll 1 .\tag{8.137}$$

For small fractional bandwidth $\Delta f/f_0$ the form of the main lobe is independent of the signal form as shown by the derivation of equation (8.137).

Now consider an array whose element spacing is $D > \lambda_0$. Assuming that the delay between the arrival of the signal at two adjacent elements is longer than the pulse duration: $\tau = T$ or

$$\Delta f/f_0 = 1 .\tag{8.138}$$

Here the influence of the modulation function $u(t)$ on the array characteristic can no longer be neglected. The secondary lobes are attenuated. DiFranco and Rubin (61) have indicated a way to take advantage of the relation between signal form and antenna characteristic to reduce the number of sidelobes of an N-element interferometer by decreasing their amplitudes to the $1/N$-fold of the main lobe amplitude. This procedure is discussed in the next section.

8.2.4.5. Ambiguity Function of Radar Systems with a Linear Antenna Array or Linear Interferometer

8.2.4.5.1. Signal bandwidth and ambiguity We consider a linear array of length L consisting of elements fed in phase and oriented parallel to the earth's surface. If echo signals are normally incident on this aperture, i.e. from the *zenith*, then the energy is in phase as all elements are reached at the same time. For a deviation from the zenith of an angle ϑ a difference of arrival time occurs (equation (8.134)) between the two most distant elements of this aperture.

Further the angular difference between the zenith angle ϑ of the target ("target angle") and the zenith angle ϑ^* of the central axis of a grating lobe of the array pattern may be defined. It is usually appropriate to use the following modified expressions (Figs. 8.3, 8.4)

$$\xi = \sin \vartheta , \quad \xi^* = \sin \vartheta^* . \tag{8.139}$$

As the voltage available at the output terminals of the array is a function of the direction parameters ξ and ξ^* there exists also a dependence on ξ and ξ^* of the echo signal to be processed which will be written as $s(t, \xi, \xi^*)$. The complex a.f. is thus also a function of ξ and ξ^* (218)

$$\chi(\tau, f_d, \xi, \xi^*) = \int\limits_{-\infty}^{+\infty} u(t, \xi, \xi^*) \, u^*(t - \tau, \xi, \xi^*) \, e^{j2\pi f_d t} \, dt . \tag{8.140}$$

In most cases to be considered the echo signal is only dependent on the difference $\Delta\xi = \xi - \xi^*$, so that $s(t, \xi, \xi^*)$ can be changed to $s(t, \Delta\xi)$ and the a.f. to $\chi(\tau, f_d, \Delta\xi)$.

Now a system shall be considered in which a stationary transmitting antenna illuminates an area to be surveyed and the linear array interferometer system is used for the reception of the echo signals reflected by the target. The system shown in Fig. 8.18 is provided with delay units at the outputs of the antenna elements in order to generate radiation lobes at several zenith angles ϑ. The amplitude taper p_n of the n-th array element is assumed purely real.

Now the a.f. must be determined from the output function of the matched filter bank associated with the lobe axis which is oriented toward the zenith. The target angle ϑ can, however, take on several different values. If we investigate the output function of a *matched filter bank* belonging to a lobe which is away from the zenith, the same result is obtained. If the target is positioned exactly in the zenith, i.e. $\Delta\xi = 0$ then the signal is in phase at all antenna outputs and the signal which appears at the output of the combiner corresponds to the signal radiated by the transmitter but delayed by the propagation time τ and distorted by the Doppler effect.

The matched filters at the output of the combiner are tuned to delayed and frequency-shifted signals, so the two dimensional a.f. for the signal concerned is available at the output of the matched filter bank. The a.f. has this form shown in Fig. 8.20a which although optimal is not realizable in practice. For another case in which the target is distant by the angle ϑ_1 from the zenith axis, i.e. from the normal to the array aperture, the a.f. is shown in Fig. 8.25.

Here assume that the pulse duration i.e. of a pulse modulated for example by *phase switching* is much shorter than the time difference between the incidence of two signals at two adjacent elements so that a temporal overlapping of two pulses processed by two different array elements is impossible. A uniform illumination of all elements is assumed. The time between two peaks of the a.f. (Fig. 8.25) can be derived as

$$\tau = (D/c)\Delta\xi_1 . \tag{8.141}$$

Fig. 8.26 shows a.f. of τ and $\Delta\xi$ for $f_d = 0$ in a presentation equivalent to Fig. 8.25. Normally, the equation for the radiation pattern is a function of the element spacing, where for

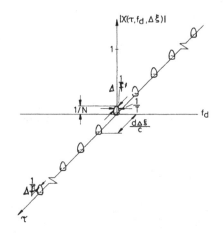

Fig. 8.25.   *Ambiguity function of broadband*
signals for the target position
outside of the zenith (218,221,222).

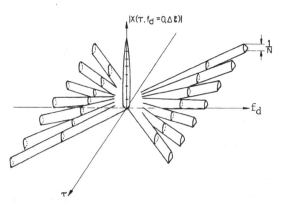

Fig. 8.26.   *Ambiguity function of broadband*
signals in the $\tau - f_d$-plane
(218).

$$k = D/\lambda_0 > 1 \tag{8.142}$$

grating lobes appear, where k is the baseline (see (8.24)) expressed in wavelengths of the carrier. This dependence is not evident from Fig. 8.26. As shown above, the normalized a.f. has a maximum amplitude $= 1$ if the target position lies exactly in the zenith. This a.f. for which the statements of section 8.2.4.2 are valid, depends only on the signal used. For target angles deviating from the zenith, i.e. for $\Delta \xi \neq 0$, the signals received by the individual antennas are no longer in phase and cannot be superposed.

Thus every single array element produces a *normalized a.f.* of height $1/N$. The element spacing D defines only the time spacing at which these N a.f. occur whose amplitudes are independent of D. The assumed short pulse duration requires a broad signal bandwidth. Now the question arises how the a.f. behaves for narrow-banded signals, i.e. for $\Delta\xi \neq 0$ the duration of the compressed pulse appearing at the matched filter are considerably longer than the difference between the arrival times at two adjacent elements, so that the envelopes of the pulses received at the elements actually overlap. The carriers also show a relative phase shift. The response of a matched filter to a signal coming from the n-th array element corresponds then to the *crosscorrelation* of this signal with the reference signal multiplied by p_n and phase shifted by $a = \Delta\xi d\omega_0 n/c_0$ (analogous to equation (8.81))

$$\chi(\tau, f_d, 0)\, p_n\, e^{j\Delta\xi d\omega_0 n/c_0} \; .$$

The output function of a matched filter for the whole array is then

$$s_0(\tau, f_d) = \chi(\tau, f_d, 0) \sum_{n=1}^{N} p_n\, exp(j\Delta\xi d\omega_0 n/c) \; . \tag{8.143}$$

The p_n are normalized

$$\sum_{n=1}^{N} p_n = 1 \; . \tag{8.144}$$

With $p_n = 1/N$ and the rules for geometric sums, the envelope of the matched filter function yields

$$|s_0(\tau, f_d)| = (1/N)\,|\chi(\tau, f_d, 0)|\, \frac{\sin N \dfrac{\Delta\xi d\omega_0}{2c}}{\sin \dfrac{\Delta\xi d\omega_0}{2c}} \; . \tag{8.145}$$

For completeness the *last* term in equation (8.145) is derived here

$$\sum_{n=1}^{N} exp(jan) = \frac{exp\big(ja(N+1)\big) - 1}{exp(ja) - 1} - 1 =$$

$$= \frac{exp(j(N+1)a/2)}{exp(jNa/2)}\, \frac{exp(j(N+1)a/2) - exp(-j(N+1)a/2)}{exp(ja/2) - exp(-ja/2)} - 1 =$$

$$= exp(jNa/2)\, \frac{\sin(N+1)a/2}{\sin(a/2)} - 1 =$$

$$= exp(jNa/2)X - 1 \tag{8.146}$$

where

$$X = \frac{\sin(N+1)a/2}{\sin a/2} \; . \tag{8.147}$$

Thus the absolute value is

$$\left|\sum_{n=1}^{N} e^{jan}\right| = \left(\big(X\,\cos(Na/2) - 1\big)^2 + \big(X\,\sin(Na/2)\big)^2 \right)^{1/2} =$$

$$= (X^2 - 2X \cos(N a/2) + 1)^{1/2} \tag{8.148}$$

with

$$\sin(N + 1)a/2 = \sin N a/2 \cos a/2 + \cos N a/2 \sin a/2 \tag{8.149}$$

with

$$\left|\sum_{N=1}^{N} e^{jan}\right| = \frac{1}{\sin a/2} \left(\sin^2(Na/2) \cos^2(a/2) - \cos^2(Na/2)\right.$$

$$\left.\sin^2(a/2) + \sin^2(a/2)\right)^{1/2} =$$

$$= \frac{1}{\sin a/2} \left(\sin^2(Na/2) \cos^2(a/2) - \sin^2(a/2)(1 - \cos^2(Na/2))\right)^{1/2} =$$

$$= \frac{\sin(Na/2)}{\sin(a/2)} \; . \tag{8.150}$$

From the equations (8.143) and (8.144) we may recognize that the shape of the a.f. in the $\tau - f_d$-plane is independent of the angle between target and vertical, except for the fact that amplitude of the peak is determined by the last term of equation (8.143).

Fig. 8.27 shows the a.f. in the $\tau - \Delta\xi$-plane which has a dependance on the factor

$$k = D/\lambda_0 \; , \tag{8.151}$$

since for

$$k > 1 \tag{8.152}$$

grating lobes occur, which disappear for

$$k < 1 \; . \tag{8.153}$$

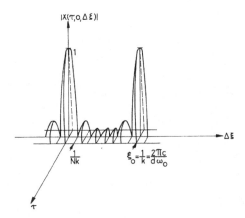

Fig. 8.27. *Ambiguity function of narrow band signals (218).*

Now we consider how to choose the bandwidth for $k > 1$ in order to reduce the number of grating lobes and so diminish the ambiguity.

We first deal with the case in which the amplitude of the grating lobes is reduced to the $1/N$-fold of the amplitude of the main lobe. This is just still correct when the modulated pulses coming from the single array elements just intersect at the *halfvoltage points* (63,68), i.e. at the points which define the pulse duration. The pulse only overlaps at the points defined when the interval between pulses arriving at the antennas is equal to the pulse duration.

This condition determines the pulse duration and thus also the signal bandwidth. At an angle of incidence determined by the direction of the axis of the main grating lobe and the normal to the aperture, i.e.

$$\Delta \xi_0 = 1/k = \lambda_0/D \tag{8.154}$$

the time between pulses becomes

$$\tau = D.\Delta \xi_0/c_0 = D/\lambda_0 f_0 k = 1/f_0 \tag{8.155}$$

for the conditions stated (Fig. 8.27). Therefore the pulse duration must be $1/f_0$ and equally the signal bandwidth f_0.

If the signal bandwidth is reduced, i.e. the pulse duration is increased then the modulated pulses intersect above the half voltage points, and an increase of the amplitude of the main grating lobe results. For a given bandwidth the required grating lobe level can also be achieved by using a sufficient number of array elements.

In conclusion it may be noted that the form of the a.f. given in Fig. 8.26 is not obtainable merely by using large antenna spacings if the bandwidth is small. Although the ambiguity function can have the form shown in Fig. 8.26 for large D and $\Delta \xi$ for narrow bandwidths grating lobes will still appear for small $\Delta \xi$.

8.2.4.5.2. Transfer function The receiving antenna can be regarded as a system with a transfer function which depends on the angle of incidence ϑ (see Fig. 5.1). Using the transfer function one can state in a simple manner the voltage produced across a load when a signal is incident on the antenna at an angle ϑ. The transfer function of the receiver (R) is written as $H_R(\omega, \xi, \xi^*)$. Since the function generally depends on $\Delta \xi$, it can be given as $H_R(\omega, \Delta \xi)$. Assuming a uniform illumination of the array antennas the transfer function is

$$H_R(\omega, \Delta \xi) = (1/N)\, \sin(N \Delta \xi d\omega/2c)/\sin(\Delta \xi d\omega/2c)\,, \tag{8.156}$$

where c is the velocity of light. The matched filter output signal is given by the inverse Fourier transform of the product of the spectrum of the incident signal with the *transfer function of the antenna array* and the matched filter.

Fig. 8.28 plots array transfer function against frequency and $\Delta \xi$ (218). This is the normal one-dimensional representation of the antenna pattern of an array with the addition of a frequency dependence. In the zenith, i.e. at $\Delta \xi$, the antenna behaves as an allpass (68).

Here the a.f. should be equal to the a.f. of the signal used. For a deviation of the target position from the zenith direction this is no longer true, as the matched filter output function is then given by the products $\chi(\omega, f_d, 0)\, H_R(\omega, \Delta \xi)$, where $\chi(\omega, f_d, 0)$ is equal to the Fourier transform of the a.f. $\chi(\omega, f_d, 0)$ appropriate to the signal in use. In order to determine the a.f. for the case of broadband signals, it is useful to express the function $H_R(\omega, \Delta \xi)$ as a geometric sum

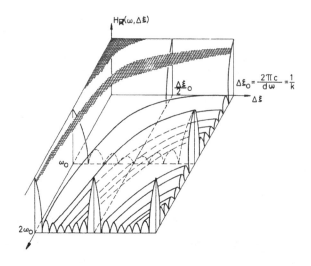

Fig. 8.28.   Transfer function of a linear
antenna array: $N = 8$ (218).

$$(1/N)\sin(N\Delta\xi d\omega/2c_0)/\sin(\Delta\xi d\omega/2c_0) = (1/N)\sum_{n=1}^{N}\exp(nj\Delta\xi d\omega/c_0) . \qquad (8.157)$$

If the equation (8.140) is applied taking into account that a multiplication by $\exp(j\omega t)$ corresponds to a delay t, the a.f. at $\Delta\xi \neq 0$ is obtained

$$\chi(\omega,f_d,0)\,H_R(\omega,\Delta\xi) \circ{-}\circ \chi(\tau,f_d,\Delta\xi)$$

$$\chi(\omega,f_d,0)(1/N)\sum_{n=1}^{N} e^{nj\Delta\xi d\omega/c} \circ{-}\circ (1/N)\sum_{n=1}^{N}\chi\left(\tau-\frac{n\Delta\xi d}{c},f_d\right) . \qquad (8.158)$$

For N odd

$$\chi(\omega,f_d,\Delta\xi) \circ{-}\circ (1/N)\sum_{n=(-N-1)/2}^{(N-1)/2}\chi\left(\tau-\frac{n\Delta\xi d}{c},f_d\right) . \qquad (8.159)$$

For the case of broadband signals the a.f. is once again given by Fig. 8.25 and 8.26. Fig. 8.27 shows the a.f. for the narrowband signals, as comparison with a cut along $\omega = 0$ of Fig. 8.28.

Fig. 8.28 clearly shows that no grating lobes are present when the signal bandwidth is very broad. Away from the frequency axis the antenna has narrowband behaviour, only a part of the energy of the input is passed through. The behaviour along the frequency axis is comparable to an "all pass" filter and the whole energy is passed through. If we assume bandwidth $\Delta f = f_0$ then Fig. 8.28 behaves as a bandpass with the bandwidth $\Delta f/N$, to that only the fraction $1/N$ of the incident energy is accepted.

Thus we expect that the amplitude of the first grating lobe will be reduced to the fraction $1/N$ as shown in Fig. 8.25 and 8.26.

For grating lobes at greater angular distances from the normal to the aperture, i.e. (2,3,4) $\Delta\xi$, the array possesses a comb filter behaviour. As the bandwidth of the peaks of the comb filter is equal to $1/N$ of their spacing, again only $1/N$ of the incident energy is passed, and the amplitude of lobes of the ambiguity function is reduced to $1/N$ of value.

If the fractional bandwith of the array transfer is fixed, it is nevertheless possible to achieve the required grating lobe level p by choosing an appropriate number of array elements, since

$$p = f_0/(N \Delta f' f_0) = 1/(\Delta f'N) \qquad (8.160)$$

and

$$\Delta f' = \Delta f/f_0 . \qquad (8.161)$$

$(f_0/N)/f_0$ corresponds to the fractional bandwidth of the array. The amplitude of the first grating lobe is now p times smaller than the main lobe, for instance if we have $\Delta f' = 0.1$ and required $p = 0.01$ (20dB) then using

$$N = 1/(p\Delta f') \qquad (8.162)$$

we obtain $N = 1000$.

This shows that for $k > 1$ ambiguities may be avoided, provided that an appropriate signal bandwidth and number of array elements is chosen.

The twin interferometer is a special case. For such an instrument the ambiguity function is of the form shown in Fig. 8.26 with two features proceeding from the origin, each with amplitude $1/N = 1/2$. Obviously, there is a very high probability of ambiguity of the distance and angle measurement in this case.

8.2.4.5.3. The ambiguity function for simultaneous operation The system analysed so far consisted of a receiving antenna formed by a linear group element with space $D > \lambda$ for which all lobes in the main beam are of equal amplitude, and a separate transmitting antenna. We now consider the use of the receiving antenna for simultaneous transmission (chapter 8.4). Using network theory and applying the transfer function of the transmitting antenna one obtains the electromagnetic far field normalized to unity, when a sine wave signal is applied to the antenna terminals. It is reasonable to normalize the transfer function in such a way that its absolute value is unity for $\Delta\xi = 0$ and $\omega = 1$. The function is written as $H_T(\omega, \Delta\xi)$. The relation connecting the transmit and receive cases is (223)

$$H_T(\omega, \Delta\xi) = j\omega H_R(\omega, \Delta\xi) . \qquad (8.163)$$

Perfect and frequency-independent matching is necessary in both cases for the validity of equation (8.163). Also, the physical extension of the array antenna must be larger than the wavelength used. On the basis of the reciprocity theorem one would expect that the transfer function in both cases would be equal but as the antenna operates in a frequency region in which geometrical optics is valid, the half-power of the main beam decreases proportional to the frequency. This decrease of lobe width is equivalent to an increase of the field strength in the axial direction, thus for a transmitting antenna this field strength is directly proportional to the frequency. On the other hand, the voltage generated at the load of a receiving antenna induced by an incident plane wave of constant intensity is frequency-independent.

When determining the output function of a matched filter the factor $j\omega$ in equation (8.163) is initially not taken into account. As the signal passes the antenna only once for each of the two cases of transmission and reception, the resulting transfer function $\{H_R(\omega, \Delta\xi)\}^2$ and the output function of the matched filter is equal to the product of $\{H_R(\omega, \Delta\xi)\}^2$ and the spectrum of the a.f. $\chi(\omega, f_d, 0)$ for the signal in use. For a broadband signal we obtain

$$\chi(\omega, f_d, 0)\, \{H_R(\omega, \Delta\xi)\}^2 \;\; \circ\!\!-\!\!\circ \;\; |\chi(\tau, f_d, 0)| \left| \sum_{n=-(N-1)}^{N-1} \frac{N-n}{N^2}\, e^{jn\Delta\xi d\omega/c} \right| =$$

$$= |\chi(\tau, f_d, \Delta\xi)| . \tag{8.164}$$

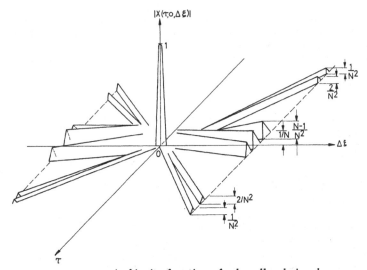

*Ambiguity function of a broadband signal
with large time-bandwidth product for
simultaneous operation of an antenna as
transmitting and receiving element (218).*

Fig. 8.29 shows the shape of the a.f. in the $\tau - \Delta\xi$-plane. If the factor $j\omega$ is included then the shape is not only changed in the $\tau - \Delta\xi$-plane but also in the $\tau - f_d$-plane. If the terms in equation (8.163) are expressed for clarity in the notation of complex modulation envelopes, then considering equation (8.163) in the time domain

$$H_T(t, \Delta\xi) = (d/dt)\, \{H_R(t, \Delta\xi)\} \tag{8.165}$$

with

$$H_T(t, \Delta\xi) = h_T(t, \Delta\xi)\, e^{j\omega_0 t}$$

$$H_R(t, \Delta\xi) = h_R(t, \Delta\xi)\, e^{j\omega_0 t}$$

$$h_T(t, \Delta\xi)\, e^{j\omega_0 t} = (d/dt)\, \{h_R(t, \Delta\xi)\, e^{j\omega_0 t}\}$$

we obtain

$$= e^{j\omega_0 t}\, (d/dt) h_R(t, \Delta\xi) + j\omega_0\, h_R(t, \Delta\xi)\, e^{j\omega_0 t} \; . \tag{8.166}$$

Because $j\omega_0 = $ constant then in the frequency domain

$$h_T(\omega, \Delta\xi) = j(\omega + \omega_0)\, h_R(\omega, \Delta\xi) =$$

$$= j\omega_0\, h_R(\omega, \Delta\xi) + j\omega\, h_E(\omega, \Delta\xi) \; . \tag{8.167}$$

In the transmitting case the transfer function of the antenna can be divided into two components, of which one except for a constant factor ω_0 is identical to the transfer function of the case in which the factor $j.\omega$ was not considered. The second component is equal to the product of $j\omega$ and the same transfer function, which in the time domain is equivalent to the first time derivative. In the same way it follows that the output function of the matched filter also consists of two parts, i.e. the a.f. is given by the absolute value of the function by the superposition of the function which appears without taking account of the factor $j\omega$ at the output of the matched filter and the first time derivative of this function. In the case of narrow band signals this differential term can be neglected. Fig. 8.29 represents the a.f. for simultaneous transmit-receive operation.

8.2.5. Influence of the Atmospherical Refraction on the Angular Measurement

Because of atmospheric refraction the elevation angle ϑ'_n at which the plane electromagnetic wave intersects the earth's surface is not identical to the elevation angle ϑ' of the line connecting the centre of the aperture and target P (Fig. 8.30).

This section discusses errors caused by the atmosphere in angular measurement using a twin interferometer (223,224,225,226). In Fig. 8.30 the geometrical relations are given for the case of an interferometer with very small baseline D or of a parabolic reflector antenna with an aperture diameter "D", where D is so small that the curvature of the earth surface can be neglected compared to D. The values of ϑ'_0 and ϑ' differ by an error

$$\varepsilon = \vartheta'_0 - \vartheta_0 \; . \tag{8.168}$$

Because a parabolic antenna may be slewed and tipped the situation shown in Fig. 8.30 remains constant for a large range of elevation angles. Assume that the rays are approximately parallel the following relation is valid

$$\cos \vartheta'_0 \approx s_0/D \tag{8.169}$$

with

$\vartheta'_0 = $ angle of incidence derived from an interferometer phase
measurement, approximately equal to ϑ'
$s_0 = $ additional path length which ray one has to traverse
compared to ray two
$D = $ baseline (inter-element distance).

For a twin interferometer with a large baseline length the conditions are different. These are now investigated, whereby the following limitations are applied.

Fig. 8.30.   *Geometry for the refraction computation omitting the curvature*
                of the earth (224).

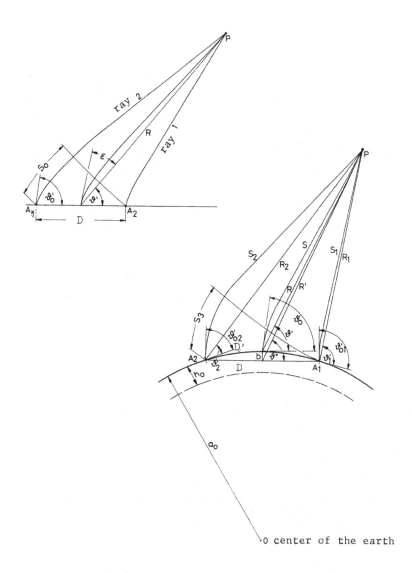

Fig. 8.31.   *Influence of the atmospheric refraction on a twin interferometer*
                (225).

1. The ratio of range R to the target P and antenna aperture satisfies the relation $R/D = 100.0$.

2. The elevation angle ϑ' is within the limits $10° < \vartheta' < 90°$.

3. The atmosphere consists of a superposition of uniform spherical layers.

If R_{o1} and R_{o2} are the distances between the individual antenna elements and the target an angle of incidence which depends on the range difference

$$\Delta R_o = R_{o2} - R_{o1} \tag{8.170}$$

is obtained as a result of the interferometer phase measurement. R_{o1} and R_{o2} are defined by the following relation

$$R_{oi} = \int_0^{s_i} n(s)ds \tag{8.171}$$

where
$n(s) =$ the index of rotation of the atmosphere as a function of the distance s along the range of total length s_i. Further an average index of refraction may be defined

$$\bar{n}_i = (1/s_i) \int_0^{s_i} n(s)ds \ , \tag{8.172}$$

from which

$$R_{oi} = \bar{n}_i \, s_i \ . \tag{8.173}$$

Equally R_{oi} can be given by the relation

$$R_{oi} = R_i + \Delta R_{oi} \ , \tag{8.174}$$

where R_i corresponds to the direct range to the target and ΔR_{oi} to the extra component of R_{oi} caused by refraction. From the equations (8.170, 173, 174) the following expressions may be derived

$$R_o \ = \ R_2 - R_1 + \Delta R_{O2} - \Delta R_{O1} = \Delta R + \delta \tag{8.175}$$

$$\Delta R = \ R_2 - R_1 \tag{8.176}$$

$$\delta \ \ = \ \Delta R_{O2} - \Delta R_{O1} \ . \tag{8.177}$$

Applying (8.173) and (8.174)

$$\delta = \bar{n}_2 s_2 - \bar{n}_1 s_1 - \Delta R \ . \tag{8.178}$$

It can be shown that

$$(\bar{n}_2 - \bar{n}_1)/\bar{n} < 1 \tag{8.179}$$

if $n(s)$ varies exponentially with height h and $\bar{n}$ corresponds to the *average index of refraction* over the distance s. Bean and Thayer (227,228,229,300) derive the following relation for n

$$n = n_0 \exp\left(-c_e(h - h_0)\right) \tag{8.180}$$

where

n_0 = index of refraction over the surface of the earth,
h = height of the target,
h_0 = height of the radar system above sea level (Fig. 8.31),
c_e = $\ln(n_0/n_1)$, where n_1 is equal to the index of
       refraction 1. km altitude.

The error of diffraction caused by earth curvature is

$$\delta R_i/s_i = (s_i - R_i)/s_i < 1 \ . \tag{8.181}$$

From equation (8.179) and (8.181)

$$\bar{n} \approx \bar{n}_1 \approx \bar{n}_2 \tag{8.182}$$

and the diffraction error may be neglected.
   With the assumptions, equation (8.178)

$$\delta = \bar{n}(\delta R_{02} + R_2 - \delta R_{01} - R_1) - \Delta R$$

$$\delta/\Delta R \approx \bar{n} - 1 \ . \tag{8.183}$$

If both rays arrive almost parallel at the interferometer elements A_1 and A_2 (Fig. 8.31) the following relation is valid in analogy to equation (8.169)

$$\cos \vartheta_0'' = s_3/D \ , \tag{8.184}$$

where ϑ_0'' is again a good approximation to ϑ_0'. As an interferometer is not tiltable like a single parabolic reflector antenna, for ray two an additional path length s_3 results which is a function of the deviation angle, and for this path the influence of refraction must also be considered (12,125) defines s_3 as

$$s_3 = \Delta R_0/\bar{n}_3 \ , \tag{8.185}$$

where

$$\bar{n}_3 \approx n_0 \ . \tag{8.186}$$

s_3 is close to the earth's surface.
   Using equation (8.175), (8.184), (8.185) and (8.186) we obtain

$$\vartheta_0'' = \text{arc } \cos(\Delta R/Dn_0)\left(1 - (\delta/\Delta R)\right) \ . \tag{8.187}$$

In order to simplify the following derivation, equation (8.187) will be modified. First equation (8.183) is substituted into (8.187) and we introduce $+1$ and -1 so that

$$\vartheta_0'' = \text{arc } \cos(\Delta R/Dn_0)(1 - \bar{n} - 1) =$$

$$= \text{arc } \cos(\Delta R/D)(\bar{n}/n_0) =$$

$$= \text{arc } \cos(\Delta R/D)(1 + (\bar{n}/n_0) - 1) \ . \tag{1.188}$$

Substituting

$$\vartheta_0'' = \text{arc cos } x(1 + y) \tag{8.189}$$

with

$$x = \Delta R/D \tag{8.190}$$

and

$$y = (\bar{n}/n_0) - 1 . \tag{8.191}$$

Then the elevation angle of the source of radiation ϑ'' may be derived using the cosine rule

$$R_1^2 = (D/2)^2 + (R')^2 - R'D \cos \vartheta'' \tag{8.192}$$

$$R_2^2 = (D/2)^2 + (R')^2 + R'D \cos \vartheta'' \tag{8.193}$$

$$\cos \vartheta'' = (R_2^2 - R_1^2)/2DR' \tag{8.194}$$

and

$$\vartheta'' = \text{arc cos}(\Delta R/D)(R_1 + R_2)/(2R') = \text{arc cos } x(1 + y') \tag{8.195}$$

where

$$1 + y' = (R_2 + R_1)/(2R') . \tag{8.196}$$

Then equation (4.20) and (4.23) are expanded as a Taylor series

$$f(x + h) = f(x) + (h/1!)f'(x) + (h^2/2!)f''(x) + ... \tag{8.197}$$

where

$$h \quad = \quad xy$$

$$f(x) \quad = \quad \text{arc cos}(x)$$

$$f'(x) \quad = \quad -1/(1 + x^2)^{1/2}$$

$$f''(x) \quad = \quad -x/(2!(1 + x^2)^{3/2})$$

$$f'''(x) = \quad -x^2/(3!(1 + x^2)^{5/2})$$

yielding

$$\vartheta_0'' = \text{arc cos}(x) - \frac{xy}{1!(1 - x^2)^{1/2}} - \frac{x^3y^2}{2!(1 - x^2)^{3/2}} - \frac{x^2 - (1 + 2x^2)}{3!(1 - x^2)^{5/2}} - ... . \tag{8.198}$$

Likewise

$$\vartheta'' = \text{arc cos}(x) - \frac{xy'}{1!(1 - x^2)^{1/2}} - \frac{3(y')^2}{2!(1 - x^2)^{1/2}} - ... . \tag{8.199}$$

Following equation (8.168) we use the difference of angles ϑ_0'' and ϑ' in order to express the *error due to refraction* expected in an interferometer measurement

$$\epsilon' = \vartheta_0'' - \vartheta'' \quad =$$

$$= \text{arc cos}(x) - (xy)/(1 - x^2)^{1/2} - ... - \text{arc cos}(x) + (xy)/(1 - x^2)^{1/2} + ... \tag{8.200}$$

The influence of y' on ϵ' is purely geometric and is only relevant if the target range is comparable to the element spacing D. Only the contribution of y to ϵ' is a consequence of atmospheric refraction and will be taken into account. Neglecting quadratic terms and all terms of higher order we obtain

$$\epsilon' = -xy/(1 - x^2)^{1/2} \tag{8.201}$$

and with the restrictions from equation (8.199)

$$\vartheta'' = \arccos(x) . \tag{8.202}$$

Introducing (8.188), (8.189), and (8.202) in (8.201) yields

$$\epsilon' = \left(1 - (\bar{n}/n_0)\right) \cos(\vartheta'')/(1 - \cos^2 \vartheta'')^{1/2} =$$

$$= \left(1 - (\bar{n}/n_0)\right) \cot(\vartheta'') \tag{8.203}$$

which corresponds to the result of Paul (224). As Bean and Thayer (227) do not take the refraction along s_3 into account they obtain the relation

$$\epsilon' = (1 - \bar{n}) \cot(\vartheta'') . \tag{8.204}$$

The difference

$$\Delta\epsilon = \epsilon' - \epsilon \tag{8.205}$$

yields a measure of the difference between a measurement made with an interferometer and with a single parabolic antenna situated midway between A_1 and A_2.

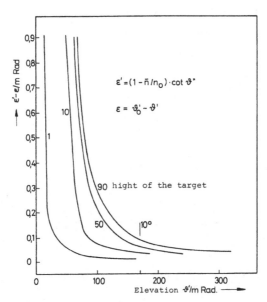

Fig. 8.32.   Comparison of the refraction error ϵ
with the refraction error of an interfero-
meter when refraction along s_3 is considered ϵ' (25).

In Fig. 8.32 the dependence of $\Delta\epsilon$ on elevation angle ϑ'' is shown. The curve shown is *valid* for $n_0 = 1,000313$; the values of the other parameters as $\bar{n}$, ϵ' and ϵ can be taken from tables (225).

A further error in the computation of the error due to refraction stems from the fact that the curved baseline D' is normally replaced by its chord (Fig. 8.31)

$$D' = 2\,a_0\,\text{arc}\,\sin(D/2a_0)\,,\tag{8.206}$$

with a_0 = geocentric radius.

But this error is for the example of a curved baseline of 5. km so small that it can be neglected. It should be mentioned that knowledge of ϵ' is not important for radio astronomers who generally observe well above the horizon as for radar observers, who frequently observe objects at low elevation angles.

8.3. Two-Dimensional Phase-Comparison Monopulse Radar

Fig. 8.33 shows the block diagram of a radar operating on the interferometer principles (70,76,81,233). One of the two parabolic antenna elements is provided with a *duplexer* for transmission and reception while the other antenna element has only a receive channel. Both receiver channels are identical.

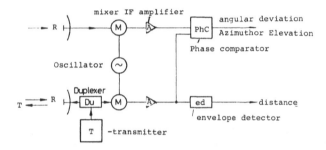

Fig. 8.33.   Two dimensional monopulse radar with phase comparison.

The radio frequency (RF) target echo signals are transformed to intermediate frequency (IF) signals by mixing with a local oscillator (LO). The output signals of both IF amplifiers are connected to a phase comparison circuit whose output voltage is proportional to phase, where $\Phi = (2\pi D \sin\vartheta)/\lambda_0$ (6.1). This voltage corresponds to an angle error, and can be applied to a control circuit which actuates a motor to move the antennas into such a position that the error is cancelled. This is the case when the target lies along the azimuth axis. One of the receiving channels contains an *envelope detector* which can be used to extract range information from the echo signal as in a normal radar. Compared to conventional systems this system has the disadvantage that the sidelobe alternation is lower than that of a single parabolic reflector, and that the available antenna aperture is not fully utilized. As both antennas are used for reception but only one for transmission the antenna effective area or gain introduced into *the radar equation* (32,70,81,86) is only that of a single antenna

$$R_{\max} = \left\{ P_T\,G\,A_e\,\sigma / \left((4\pi)^2\,P_{\min}\right) \right\}^{1/4}\,.\tag{8.207}$$

The parameters in this equation are:

$R_{\max}$ = maximum radar range (m)
P_T = radiated transmitter power (W)
A_e = antenna area (m^2)
G = radiative gain
σ = radar cross-section of the target
        = 4π. power reflected back to transmitter per unit
        solid angle divided by the power incident on the
        target surface unit
$P_{\min}$ = minimum received power.

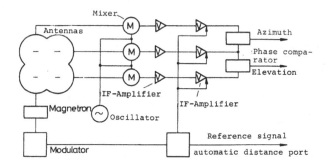

*Fig. 8.34a.   Two planes monopulse system with
phase comparison.*

8.4. Phase-Comparison Monopulse Radar for Three-Dimensional Target Tracking

This section describes an interferometer radar system capable of tracking in two planes. The block diagram of a 3-cm system is shown in Fig. 7.35. Four parabolic reflectors of 0.4 m diameter are spaced by 0.38 m. One element is used for transmission while the remaining three are arranged so that two form an interferometer in elevation and two are sensitive in the azimuth direction, one reflector being used in both planes. The received signals are mixed with a common LO and amplified in IF channels, the relative amplitudes and phases of the RF signals being preserved. The mixers, oscillators and IF amplifiers are located at the back of the *cloverleaf antenna*, the other components are on a fixed base and are connected to the antenna via a flexible cable. A single-channel rotary joint is used to feed the transmit signal from the *magnetron* to the antenna. An automatic gain control system prevents amplitude variations at the IF output. An automatic *range gate* is automatically synchronized by the modulator to track the distance of the target. The two pairs of constant-amplitude IF signals are applied to *phase comparators* which use quadratic detectors. The output of the comparator is zero when the target is on the axis of the system.

Above and to the right of the main axis these error signals are positive below and to the left of the main they are negative. After the amplification the *video pulses* are smoothed and used to drive a motor which moves the antenna so that the target is centred. The monopulse receiver and the antenna drive form a closed *servoloop-mechanism*. When the target crosses the main axis, the smoothed output of the receiver changes its polarity and the rotation direction of the motor is reversed. In this way automatic *three-dimensional tracking* in azimuth and elevation is achieved. This system can be upgraded to give better resolution by increasing the diameter of the reflectors. It is also possible to make all four reflectors radiate with equal power and with equal phase in order to achieve maximum target illumination. For a simple four-horn-*monopulse-feed* the aperture centers of the four hours are at the corners of a square with sides of length a. The formula

$$a/\lambda_0 = f_p/D_p ,\qquad\qquad\qquad(8.208)$$

where f_p and D_p are the parabola focal length and diameter enables the choice of optimum antenna parameters for a centre frequency. The drawback of the monopulse principle lies in the uncertainty in identifying the correct target if several echo signals of equal intensity from several targets are present.

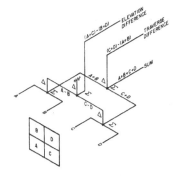 

Fig. 8.34.   *Two planes monopulse system*
  b. Block diagram for sum and difference signal;
  c. Cloverleaf antenna (General Electric Co. (70, p. 79,91).

8.5. Pseudo-Noise Code Interferometer

An integrated system will now be described consisting of a *transponder* on board of an aircraft which radiates signals modulated by a pseudo-noise code or a pseudo-random code modulated by a pseudo-random code modulated *subcarrier* together with a twin interferometer for the reception of these signals (224). Fig. 8.35 shows the geometry of this system. The *bearing* ϑ' of the target can be derived by application of the cosine

$$R_2^2 = (D/2)^2 + R^2 - (D/2)R\,2\cos\vartheta' = D^2/4 + R^2 - DR\cos\vartheta'$$

$$R_1^2 = (D/2)^2 + R^2 - 2R(D/2)\cos(180° - \vartheta') .\qquad\qquad(8.209)$$

$$\cos(180° - \vartheta') = -\cos\vartheta'\qquad\qquad\qquad(8.210)$$

then

$$R_1^2 = D^2/4 + R^2 + R\,D\cos\vartheta' .\qquad\qquad\qquad(8.211)$$

By elimination of $\cos\vartheta'$ in (8.209) and (8.211)

$$R_1^2 + R_2^2 = D^2/2 + 2R^2 \ . \tag{8.212}$$

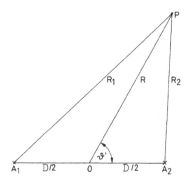

Fig. 8.35. Geometry of a radar twin interferometer.

If (8.209) is solved for $\cos \vartheta'$

$$\cos \vartheta' = (R_2^2 - (D^2/4) - R^2)/(-RD) = \left(R^2 - R_2^2 + (D^2/4)\right)/RD \tag{8.213}$$

and if (8.212) is in the result,

$$\cos \vartheta' = \left\{(1/2)\left(R_1^2 + R_2^2 - (D^2/2)\right) - R_2^2 + D^2/4\right\}/$$
$$\left\{\pm D(1/2)\left(R_1^2 + R_2^2 - (D^2/2)\right)^{1/2}\right\} =$$
$$= \left((R_1^2/2) - (R_2^2/2)/(\pm (D/\sqrt{2})\left(R_1^2 + R_2^2 - (D^2/2)\right)^{1/2}\right.$$

$$\cos \vartheta' = \pm(R_1^2 - R_2^2)/(\sqrt{2}\,D)\left(R_1^2 + R_2^2 - (D^2/2)\right)^{1/2} \ . \tag{8.214}$$

In this derivation R_1 and R_2 are the distances between the transmitter and the receiving interferometer elements A_1 and A_2. Equation (8.214) is the exact relation for the bearing ϑ'.

This equation may be simplified by assuming that the antennas are in the far field of the radiation pattern of the radiation source. The limit between far and near field is defined by the *far field condition* (also: *Fraunhofer condition*) of a receiving aperture antenna with the largest dimension d_A (225,235).

$$R_F = 2\,d_A^2/\lambda_0 \ . \tag{8.215}$$

An additional requirement is a pointlike transmitting antenna ($d_T \ll \lambda_0$). If $d_T \gg \lambda_0$ instead of d_A^2 the expression $(d_A + d_T)^2$ has to be in (8.215). If an aircraft is positioned a distance R_F or more from the interferometer, the incident radiation may be regarded as a plane wave front and equation (8.214) can be simplified to the following expression

$$\cos \vartheta' = c_0\,\tau/d_A \tag{8.216}$$

or using

$$90° - \vartheta' = \vartheta \tag{8.217}$$

we obtain

$$\sin \vartheta = c\tau/d_A = \Phi\lambda_0/(2\pi d_A) , \tag{8.218}$$

where Φ is the phase difference and τ is the time delay between the received signals at the antennas A_2 and A_1. If now the *transponder* transmits a signal which is *phase-shift keyed* with a binary *pseudo-random code*, the time delay between the two elements can be obtained at the two matched filter outputs (Fig. 8.36) by simply measuring the time difference between the occurrence of the two a.c.f. spikes.

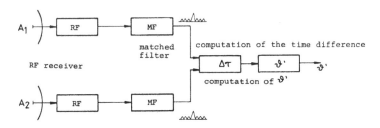

Fig. 8.36.   Block diagram of an interferometer operating with pseudo-noise-coded signals (234).

It is possible to use a subcarrier in addition to the phase modulation of carrier (75). The pseudo-random-coding allows coarse time difference determination as decribed above while the measurement of the phase differences of the received subcarriers allows a fine range determination. For the case of pseudo-noise-code modulation with an additional subcarrier the signal function (234)

$$s(t) = \cos(\omega_0 t + \Delta\kappa\varphi(t) + \Delta\Phi \cos \omega_{pi}t) \tag{8.219}$$

where

$\varphi(t)$ = pseudo-noise code (-1,+1,-1,-1,...)
ω_{pi} = angular frequency
$\Delta\Phi$ = modulation index of the subcarrier
$\Delta\kappa$ = modulation index of the pseudo-noise code
ω_0 = angular frequency of the carrier.

In order to determine the accuracy of determination of the bearing ϑ' the differential of the equation (8.216) must be formed

$$\frac{\partial}{\partial\vartheta'} (\cos \vartheta')d\vartheta' = \frac{\partial}{\partial\tau} \left(\frac{c\tau}{D^*}\right) d\tau + \frac{\partial}{\partial D} \left(\frac{c\tau}{D^*}\right) dD^* - \sin \vartheta'd\vartheta' =$$

$$= \frac{c}{D^*} d\tau + (-) \frac{c\tau}{D^*} dD^*$$

$$d\vartheta' = -\frac{c}{D^* \sin \vartheta'} d\tau + \frac{c\tau}{D^{*2} \sin \vartheta'} dD^* . \tag{8.220}$$

As the element spacing or baseline D^* generally can be regarded as constant, the second term on the right side of equation (8.220) can be set to zero. For small errors the assumption

$$d\vartheta' \approx \Delta\vartheta' \tag{8.221}$$

can be made. Then

$$\Delta T = d\tau \tag{8.222}$$

is defined as *time tolerance*. As only the absolute value in equation (8.220) is of interest the following expression is obtained

$$\Delta\vartheta' = |c/(D^* \sin \vartheta')| \Delta T \tag{8.223}$$

or

$$\Delta T = |D^* \sin \vartheta'/c| \Delta\vartheta' . \tag{8.224}$$

ϑ'	$0°$	$15°$	$30°$	$45°$	$84°$	$90°$
$\cos\vartheta'$	$1,0$	$0,966$	$0,866$	$0,707$	$0,105$	0
$\tau = \dfrac{\cos\vartheta'}{10^7}$ s	10^{-7}	$0,966 \cdot 10^{-7}$	$0,866 \cdot 10^{-7}$	$0,707 \cdot 10^{-7}$	$0,105 \cdot 10^{-7}$	0
$\sin\vartheta'$	0	$0,259$	$0,500$	$0,707$	$0,995$	$1,000$
$A = \dfrac{c}{D^* \sin\vartheta'}$		$3,860 \cdot 10^7$	$2,000 \cdot 10^7$	$1,414 \cdot 10^7$	$1,005 \cdot 10^7$	$1,000 \cdot 10^7$
$\Delta T =$ $= \dfrac{\Delta\vartheta'}{A}$ s for $\Delta\vartheta' =$ $4,5 \cdot 10^{-5}$ rad	0	$1,17 \cdot 10^{-12}$	$2,25 \cdot 10^{-12}$	$3,18 \cdot 10^{-12}$	$4,5 \cdot 10^{-12}$	$4,5 \cdot 10^{-12}$
and $1,1 \cdot 10^{-3}$ rad	0	$0,285 \cdot 10^{-10}$	$0,550 \cdot 10^{-10}$	$0,778 \cdot 10^{-10}$	$1,1 \cdot 10^{-10}$	$1,1 \cdot 10^{-10}$

Table 8.1.   Time tolerance (equation 8.224) as a function
of elevation angle.

Table 8.1 gives values of this time tolerance calculated on the basis that a given range of angular error $\Delta\vartheta'$ at $D^* = 30.$ m is not exceeded.

Two typical cases are considered and the values of $\Delta\vartheta'$ tabulated. The first case requires $\Delta\vartheta' = 4.5.10^{-5}$ rad which corresponds to a measurement accuracy of 1.6 km at 36000. km height of a geosynchronous satellite. In the second instance $\Delta\vartheta' = 1.1.10^{-3}$ rad. allows 100. m precision in the determination of the position of an aircraft flying at 10. km altitude past a beacon transmitter which is 100. km distant. The elevation angle is 6° in this case.

Various figures for the necessary bandwidth can be derived. If the carrier is modulated with a pseudo-random code the bandwidth is obtained from the *product of time and bandwidth* and is the reciprocal of the length of one binary code element

$$\Delta f = N/T \ . \tag{8.225}$$

For $T/N = 1\mu s$ we obtain $\Delta f = 1$ MHz.

If on the other hand the output of the matched filter is sampled in steps of 1 MHz, then a peak in the acf can be detected only once every 10^{-6}s, and discrimination between two acf peaks, i.e. determination of τ, is only possible with an accuracy of 2.10^{-6}s. If on the other hand the recieved signal is sampled at 10. MHz, the occurrence of an acf peak can be detected with 0.1 μs, and a double peak can be detected as such down to a spacing of 0.2 μs. Table 8.1 gives the accuracy ΔT required in the determination of τ. A value of ΔT of 10^{-12}s requires that we determine the time of an acf peak to an accuracy of $0.5.10^{-12}$s. Increasing the bandwidth to 2.10^{12} Hz would in principle satisfy the requirement but is difficult in practice.

It is, however, possible to use a bandwidth of 10^7 Hz and use a *sampling frequency* of 10^{10} Hz together with a subcarrier phase measurement. However, in practice this high sampling frequency is not possible and the system described cannot give the accuracy required to determine the position of a geosynchronous satellite. An improvement in accuracy can only be obtained by measuring the relative phases of the carrier signals received by the antennas. Thus the time measurement of the acf peaks produces a coarse position determination, the subcarrier phase refines this further in order that an accurate bearing can finally be obtained from the carrier. For instance consider a $1.5.10^9$ Hz carrier with a 10^7 Hz subcarrier. The required angular resolution of $1.8°$ corresponds to a *time resolution* (234) of

$$\Delta T = (1/200)(s/1.5 \times 10^9) = (10^{-11}/3)s = 3.33 \times 10^{-12}s \ ,$$

since $1.8°$ is $1/200$ th resolution. The length of a code element is $10^{-7}s = 0.1$ μs, corresponding to a bandwidth of 10. MHz.

k	$\sin\vartheta' = \frac{1}{2k}$	ϑ'
1	0,5	30°
1,25	0,4	23,6°
1,5	0,333	19,5°
2,0	0,25	14,5°
3,33	0,15	8,7°
4,0	0,125	7,2°
6,0	0,0834	4,75°
8,0	0,0626	3,5°
10,0	0,05	2,8°
12,0	0,0416	2,4°
14,0	0,0358	2,0°
16,0	0,0278	1,8°

Table 8.2. k-factor as a function of elevation angle (ϑ').

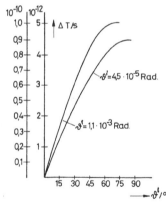

Fig. 8.37. Graph belonging to Table 8.1.

8.6. Interferometer Systems for Synchronous Satellite Navigation

The high accuracy which can be obtained using angular measurement by an interferometer has led to the use of this method for *position determination* of geo-synchronous satellites and for the alignment of the communication antennas used. Fig. 8.38 shows the geometry of such a system.

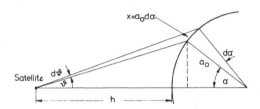

*Fig. 8.38. Geometry to finding the position of
a synchronous satellite (236).*

The symbols used are:

ϑ = bearing angle,
$d\vartheta$ = bearing error,
x = $a_0\, da$ = geodetic error for $\vartheta \neq 0$,
x_0 = geodetic error for $\vartheta = 0$,
h = height of a synchronous satellite = 36000. km =
        22500. miles,
a_0 = earth radius = 6370 km.

With these parameters the following relations may be formulated.

$$\tan\vartheta = a_0 \sin\alpha/(h + a_0 - a_0\cos\alpha) = \sin\alpha/(a - \cos\alpha) \tag{8.226}$$

where

$$a = (h + R)/R \approx 6.62 \tag{8.227}$$

α = angle between the lines connecting earth centre and inter-
        ferometer and earth centre and satellite

and

$$x/x_0 = (a^2 - 2a\cos\alpha + 1)/(a - 1)(a\cos\alpha - 1). \tag{8.228}$$

The *directional accuracy* for synchronous flight altitude can be found from Fig. 8.39 and 8.40. Fig. 8.39 shows the geodetic error in miles as a linear function of small errors $\Delta\vartheta$ for a satellite which is aligned perpendicular to the earth's surface. Fig. 8.40 shows the ratio $x(\vartheta)/x_0$ as a function of ϑ for the same bearing error as Fig. 8.39.

The function is replotted on an expanded scale for the range $0 < \vartheta < 6.5°$.
    For $\vartheta = 0°$ the geodetic error for $\Delta\vartheta = 0.125°$ using Fig. 8.39 is $x_0 = 50.$ miles. For a bearing of $\vartheta = 6°$ from Fig. 8.40 a value of $x(\vartheta < 6.5°)/x_0 = 1.46$ is obtained from the plot.

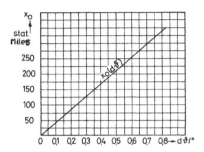

Fig. 8.39.   Geodetic error in dependence of the pointing
angle error $d\vartheta$ at $\vartheta = 0°$ (236).

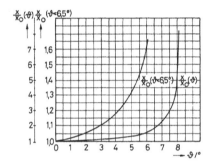

Fig. 8.40.   Normalized geodetic error as a function of the
pointing angle (236).

Assuming the same $\Delta\vartheta$ as before we derive a geodetic alignment error of $x = 1.46 \times 50 = 73$. miles. It is obvious from these values that antenna pointing must be maintained to a few tenths of a degree accuracy if a satellite carries reflector antennas with a large aperture or high directivity.

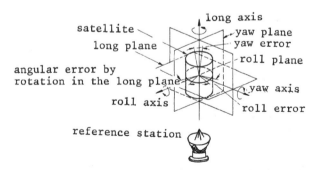

Fig. 8.41.   The stability geometry of a satellite (236).

Fig. 8.41 shows the *degrees of freedom* of a synchronous satellite. A reference station is to be tracked by an interferometer system on board of the satellite. In Fig. 8.42 the block diagram of an appropriate system is given. As the baselines of the two interferometers are orthogonal, motions around the *roll* and *yaw axis* can be measured, also the direction cosines of the reference station or of a telecommunication on the earth's surface can be determined.

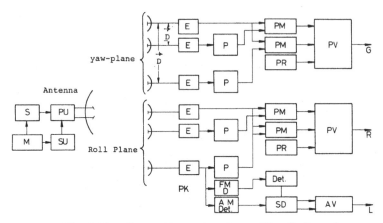

Fig. 8.42.   Cooperative system for position finding of a synchronous satellite (236).

The abbreviations in Fig. 8.42:

AV   amplitude comparison
BBS reference ground station
D   discriminator
Dec decoder
Det detector
G   angular variation around the yaw axis
L   angular variation around the pitch axis
M   modulator
P   90°-phase shifter
PK   polarization channel
PM   phase measurement
PR   phase difference
PU   polarization switching
PV   phase comparison
R   angular variation around the roll axis
S   transmitter
SN   navigation system of the satellite
SU   control for polarization switching
SD   synchronous demodulation
E   receiver.

The direction from which the signal reference station is incident is given by the "interferometer equation"

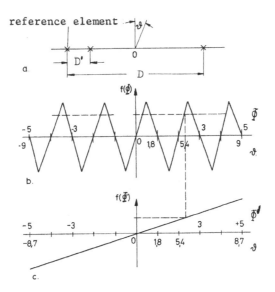

Fig. 8.43.   Resolution of the angular ambiguity by help of
a three-element interferometer.

$$\Phi = (2\pi/\lambda_0)\, D \sin \vartheta = 2\pi k \sin \vartheta \ . \tag{8.229}$$

The satellite is rotated in such a way by a control system that $\Phi = 0$ and the satellite is pointing to the reference station. The larger the phase difference Φ for small values of ϑ' the better the system operates. This is achieved in practice by using a large element spacing k. Fig. 8.43a shows the geometry of an interferometer with an element spacing or baseline of $k = 16$. The reaction of this system for a phase difference Φ, corresponding to an angle of incidence ϑ, is shown in Fig. 8.43b. It can be seen that the result is in fact rather accurate but ambiguous, i.e. every phase difference

$$(-1)^{n-1}(n\pi - \Phi)/\mathrm{rad}. \quad (n = 0, \pm1, \pm2, ...)$$

can be inferred from a measured $f(\Phi)$. The absolute value of the measured phase difference is always

$$|\Phi| = \pi - \Phi/\mathrm{rad} \ .$$

In Fig. 8.44 the corresponding ambiguous pointing angle is drawn against k for $\Phi_1 = \pi$, where for ϑ' $(n > 4)$ an expanded scale is used. The values on which Fig. 8.44 is based can be taken from Table 8.2.

It can be seen that for $k = 16$ the ambiguous angle domain begins at 1.8°, i.e. it is not known how often this angle of 1.8° must be added to the angle determined from the measured phase in order to obtain the real bearing. The measured angle ϑ is only unambiguous in the domain

$$0 < \vartheta < 1.8° \ .$$

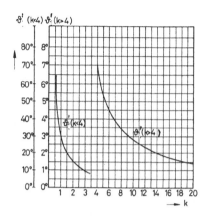

Fig. 8.44. Shape of the function of ϑ' over k (236).

The *ambiguities* occurring can be eliminated, as is well known from 7.7 and 8.2.3, by introducing a third antenna arranged on the same baseline at a distance D' from the reference antenna (Fig. 8.43a). As the bearing for a synchronous satellite can be within the angular range

$$\Delta\vartheta = 17.4° \quad \text{or} \quad \Delta\vartheta/2 = 8.7°$$

(236), the *short baseline interferometer* (SBI) is designed in such a manner that within the angular interval $0 < \vartheta < 8.7°$ only unambiguous solutions are possible, whereby it is convenient to arrange that 8.7° corresponds to a phase difference $\Phi = \pi/\text{rad}$. (Fig. 8.43c). Thus the short baseline interferometer gives a coarse but unambiguous angle measurement while the long-baseline interferometer derives a precise measurement within a small angular range. For a $\vartheta = 8.7°$ we derive $k' = 3.3$. The tolerable error of the angle can be derived from the first derivative of the interferometer equation as

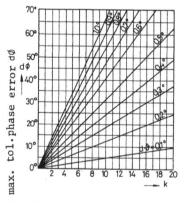

Fig. 8.45.   Maximum tolerable phase error as a function of k (236).

$$d\Phi = 2\pi k \cos \vartheta \, d\vartheta \; . \tag{8.230}$$

The maximum phase error

$$(d\Phi)_0 = 2\pi k \, d\vartheta \tag{8.231}$$

for constant k and $d\vartheta$ occurs when $\vartheta = 0°$, i.e. the satellite is exactly over the reference station. Fig. 8.45 shows the behaviour of equation (8.231). The tolerable phase error for any value of ϑ is obtained by multiplying equation (8.231) by $\cos \vartheta$.

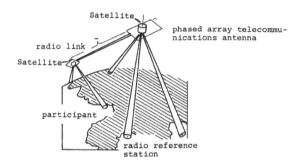

Fig. 8.46.   *Determination of position parameters with interferometer systems on board of a synchronous satellite (236).*

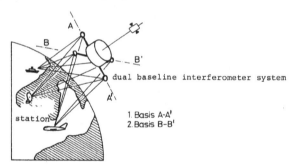

Fig. 8.47.   *Interferometer system on board of a synchronous satellite as navigation aid for aircraft and ships (236).*

Section 8.2.3 considered ambiguities from the point of view of antenna theory. However, since an interferometer baseline is large compared to wavelength a geometrical optics approach is possible. This section showed that ambiguities can be resolved for a twin interferometer by applying simple geometrical considerations.

In the case considered here a measurement of rotation about the axis of the spacecraft is possible by measuring the position angle of the linearly polarized reference signal. In this case the *Faraday rotation* (32,77), as the signal traverses the ionosphere, must be negligible. At 2 GHz this rotation is less than 2.7° and falls with $1/f^2$. Switching between two orthogonal polarizations can be used (236). In the case of microwave links this is referred to as *polarization diversity*.

The sense of rotation is determined by synchronous demodulation in the satellite receiver, while the transmitted signal must also contain synchronizing information.

Fig. 8.46 and 8.47 show possible applications. The system shown in Fig. 8.46 gives the attitude of the satellite referred to a ground station. This satellite also serves as the reference for another satellite which cannot itself receive the ground station signal. Fig. 8.47 shows an interferometer on board of a satellite which is used as a navigation aid for aircraft and ships. *Transponder* signals between the satellite interferometer system and aircraft or ships give information on distance and two direction cosines, allowing a determination of position on the earth's surface.

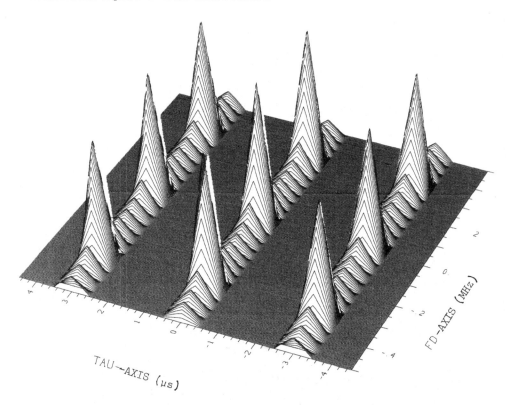

The ambiguity function (AF).

Time-shifting axis:	TAU/μs
Doppler-shifting axis:	FD/MHz
Number of pulses:	8
Pulse form:	carrier pulse
Duration of the single pulse:	0.5 μs
Pulse following time:	3. μs

(Courtesy of Dr.-Ing. P. Voigt, diss. 1990, Application of the digital correlator system DK 128 in pulse compression radars, Dept. El. Engrg., University, D-6750 Kaiserslautern, Fed. Rep. of Germany)

REFERENCES

Books (alphabetical)

Radioastronomy and Antennas

(1) Alder, B., Fernbach, S., Rotenberg, M.: 1975. Methods in computational physics, Radio Astronomy, Vol. 14, Academic Press, New York, p. 1–54, and 131–175.

(2) Bracewell, R.N.: 1959. Paris Symposium on Radio Astronomy. Stanford Univ. Press, Stanford, California, 616 p.

(3) Bracewell, R.N.: 1962. Radioastronomy techniques. Handbuch der Physik. S. Flügge, Ed. Springer, Berlin, Vol. 54, p. 42–129.

(4) Born, M., Wolf, E.: 1964. Principles of Optics. Pergamon Press, Oxford, 808 p.

(5) Cook, A.H.: 1971. Interference of electromagnetic waves. Oxford University Press, Oxford, G.B., p. 127–150.

(6) Christiansen, W.N., Högbom, J.A.: 1969. Radiotelescopes. Cambridge Univ. Press, 231 p.

(7) Christiansen, W.N.: 1973. The Fleurs Synthesis Telescope. Proceedings of the IREE (Australia), Vol. 34, No. 8, September, p. 301–355.

(8) Collin, R.E., Zucker, F.J.: 1969. Antenna Theory. Part 1, 2. McGraw-Hill, New York, Vol. I., p. 557–673.

(9) Esepkina, N.A., Korolikov, D.B., Pariskii, S.N.: 1973. Radiotelescopes and Radiometry, Iedatelistvo. Fisisko-Math. Lit., Moskva, 416 p.

(10) Evans, J.V., Hagfors, T.: 1968. Radarastronomy. McGraw-Hill, New York, 620 p.

(11) Findlay, J.W.: 1973. Radio and Radar Astronomy. Special issue of Proceedings of IEEE, September, Vol. 61, No. 9, p. 1169–1376.

(12) Grosskopf, J.: 1970. Wellenausbreitung I, II. Bibliograph. Institut, Mannheim, 215 and 472 p.

(13) Hachenberg, O., Vowinkel, B.: 1982. Technische Grundlagen der Radioastronomie. Bibliograph. Institut, Mannheim/FRG., 307 p.

(14) Hanbury-Brown, R.: 1974. The Intensity Interferometer. Taylor and Francis, London, 184 p.

(15) Hansen, R.C.: 1964. Microwave Scanning Antennas. Vol. I. Apertures. Academic Press, New York, p. 1–105.

(16) Hansen, R.C.: 1966. Microwave Scanning Antennas. Vol. II. Array and Practices. Academic Press, New York, 400 p.

(17) Hansen, R.C.: 1966. Microwave Scanning Antennas. Vol. III. Array Systems. Academic Press, New York, 420 p.

(18) Heilmann, A.: 1970. Antennen I, II, III. Bibliograph. Institut. Mannheim, 146, 167, 219 p.

(19) Hey, J.S.: 1971. The Radio Universe. Pergamon Press, Oxford, 280 p.

(20) Hoffmann, W.C.: 1960. Proceedings of the Symposium on Statistical Methods in Radio Wave Propagation. Pergamon, London, 200 p.

(21) Johnson, R.C., Jasik, H.: 1984. Antenna Engineering Handbook. McGraw-Hill, New York, p. 1600.

(22) Klinger, H.H.: 1966. Mikrowellen, Kinotechnik, Berlin, p. 149– 167.

(23) Kraus, J.D.: 1988. Antennas. McGraw-Hill, New York, 892 p.

(24) Kraus, J.D.: 1986. Radioastron. McGraw-Hill, New York, 481 p.

(25) Krug, E.: 1962. Radioastronomie. Kosmosbibliothek Frankh, Stuttgart, Band 233, 80 p.

(26) Kuehn, R.: 1964. Mikrowellenantennen. VEB Technik, Berlin. 712 p.

(27) Kuzmin, A.D., Salomonovich, A.: 1966. Radioastronomical Methods of Antenna Measurements. Academic Press, New York, 182 p.

(28) Lovell, B., Clegg, J.A.: 1952. Radio Astronomy, Chapmann Hall, London, 238 p.

(29) Mar, J.W., Liebowitz, H.: 1969. Structures Technolgy for Large Radio and Radar Telescope Systems. The MIT Press, Cambridge (Mass, USA), 538 p.

(30) Meeks, M.L.: 1976. Astrophysics, Part B: Radio Telescopes. Academic Press, New York, p. 1–118.

(31) Meeks, M.L.: 1976. Astrophysics, Part C: Radio Observations. Academic Press, New York, 1307 p.

(32) Meinke, H., Grundlach, F.W.: 1975. Taschenbuch der Hochfrequenztechnik. Springer, Berlin, 1670 p.

(33) Mikhailov, A.A.: 1967. Radio Astronomy. IPST, Jerusalem, Israel, p. 106–121, and p. 177–194.

(34) Oort, J.H.: 1974. The Synthesis Radio Telescope at Westerbork, Netherlands Foundation f. Radioastronomy, Leiden, 8 papers of Astronomy & Astrophysics 1973/74.

(35) Papas, C.H.: 1965. Theory of Electromagnetic Wave Propagation. McGraw-Hill, New York, 244 p.

(36) Pawsey, J.L., Bracewell, R.N.: 1955. Radio Astronomy. Clarendon, Oxford (G.B.), 372 p.

(37) Penfield, P., Rafuse, R.P.: 1962. Varactor-Applications. MIT-Press, Cambridge, Mass., p. 210–223.

(38) Roennaeng, B.O.: 1968. On the Theory, Techniques, and Data Processing of Very Long Baseline Interferometry. Research Rep. No. 105, Re. Lab. of Electronics, Chalmers Univ. of Technology, Gothenburg/Sweden, p. 1–59.

(39) Rusch, W.V.T., Potter, P.D.: 1970. Analysis of Reflector Antennas. Academic Press, New York, 180 p.

(40) Sahler, H., Zocher, E., Wohlleben, R.: 1968. Antennen fuer elliptisch polarisierte Wellen und ihre Messtechnik, Forschungsberichte des Landes Nordrhein-Westfalen, Nr. 1873, Westdt. Verl., Köln, p. 1–150.

(41) Sautter, H.: 1972. Astrophysik II. UTB Fischer, Stuttgart, p. 1–24.

(42) Skobel'Tsyn, D.V.: 1966. Radiotelescopes. Plenum, London, 171 p.

(43) Steel, W.H.: 1983. Interferometry. Cambridge Univ. Press, Cambridge/GB, p. 72–118, 272–280.

(44) Steinberg, J.L., Lequeux, J.: 1964. Radioastronomy. McGraw-Hill, New York, 253 p.

(45) Smith, F.G.: 1974. Radio Astronomy. Pinguin Book, Hammadsworthy/GB, 270 p.

(46) Verschuur, G.L., Kellermann, K.I.: 1974. Galactic and extragalactic radio astronomy. Springer, Berlin, p. 256–290 and p. 320–352.

(47) Verschuur, G.L.: 1974. The invisible Universe. Springer, Berlin, 170 p.

(48) Wellmann, P.: 1957. Radioastronomie. Dalph-Taschenbücher, München, Band 340, 139 p.

(49) Williams, W.E.: 1950. Applications of Interferometry. Methuen, London, 104 p.

(50) Wohlleben, R., Pfaff, K.: 1970. Fernfeldsimulation linearer Punktstrahlergruppen auf dem Analogrechner. Forschungsbericht d. Landes NRW No. 2121. Westdt. Verlag Leverkusen 3, 40 p.

(51) Kaplan, S.A.: 1971. Extraterrestral Civilisations. IPST, Jerusalem, Israel, p. 66–67.

(52) Beran, M.J., Parrent, G.B.: 1964. Theory of Partial Coherence. Prentice Hall, Englewood Cliffs, N.J., USA, 193 p.

Radartechniques, Telecommunications, Physical Interferometry

(53) Bachmann, H.W., Wasiljeff, A.: 1975. Principles of Signal Theory. Saclant ASW Res. Centre Memo. No. SM-76, I 19026 La Spezia/Italy, Dec., 67 p.

(54) Barton, D.K.: 1964. Radar Systems analysing. Prentice Hall, Englewood Cliffs (New Jersey), 350 p.

(55) Bendat, J.S.: 1958. Principles and applications of random noise theory. Wiley, New York, 270 p.

(56) Berkowitz, R.S.: 1965. Modern Radar. Wiley, New York, 380 p.

(57) Blackband, W.T.: 1967. Radio Antennas for Aircraft and Aerospace Vehicles. Technivision Services, Maidenhead (G.B.), 403 p.

(58) Burdic, W.S.: 1968. Radar Signal Analysis. Prentice Hall, p. 1–324.

(59) Candler, C.: 1970. Modern Interferometers. Hilger, London, 500 p.

(60) Cook, E.C., Bernfield, M.: 1967. Radar Signals. Academic Press, New York, 290 p.

(61) Difranco, J.V., Rubin, W.L.: 1968. Radar Detection. Prentice Hall, Englewood Cliffs (New Jersey), 646 p.

(62) Ficchi, R.F.: 1964. Electrical Interference. Iliffe, London, 262 p.

(63) Fischer, F.A.: 1969. Einführung in die statistische Übertragungstheorie. Bibliograph. Inst. Mannheim, Bd. 130/130a, p. 78–84.

(64) Herger, R.O.: 1970. Synthetic Aperture Radar Systems. Academic Press, New York, 290 p.

(65) Herriott, H.: 1967. Some Applications of Lasers to Interferometry (Progress in Optics, Vol. 6). North-Holland, Amsterdam, 205 p, (Chap. 5).

(66) IEE (G.B.): 1968. Conference on interference problems associated with the operation of microwave communication systems, IEE (G.B.). El.-Division, London, 176 p.

(67) Lange, P.H.: 1962. Korrelationselektronik. VEB Technik, Berlin, 230 p.

(68) Lueg, H. (Editor): 1970. Grundlegende Systeme, Netzwerke und Schaltungen der Impulstechnik. Lecture Notes of Institut für Technische Elektronik, TH Aachen, p. 4–48.

(69) Mönch, G.C.: 1966. Interferenzlängenmessung und Brechzahlbestimmung. Teubner, Leipzig, 260 p.

(70) Sherman, S.M.: 1984. Monopulse Principles and Techniques. Artechn House, Dedham/Mass. USA, 363 p.

(71) Neidhardt, P.: 1964. Informationstheorie und automatische Informationsverarbeitung. Berliner Union, Stuttgart, 268 p.

(72) Papoulis, A.: 1962. Systems and Transforms with Applications in Optics. McGraw-Hill, New York, 352 p.

(73) Papoulis, A.: 1962. The Fourier Integral and its Applications, McGraw-Hill, New York, 223 p.

(74) Peters, J.: 1967. Einführung in die allgemeine Informationstheorie. Springer, Berlin, 266 p.

(75) Rihaczek, A.: 1969. Principles of High-Resolution Radar. McGraw-Hill, New York, p. 1–225.

(76) Rhodes, D.R.: 1959. Introduction to Monopulse. McGraw-Hill, New York, 13 p., (119 p.), or Artechn House, Dedham/Mass. USA, 1980, 119 p.

(77) Skolnik, M.I.: 1962. Introduction to Radar Systems, McGraw-Hill-ISE, New York, 648 p.

(78) Skolnik, M.I.: 1970. Radar Handbook, McGraw-Hill, New York, 1740 p.

(79) VDE-NTG-URSI: 1968. Antennen und elektromagnetische Felder. VDE-Bezirksverein, Frankfurt, 73 p.

(80) Rautenfeld, F. von: 1957. Impulsfreie elektrische Rueckstrahlverfahren (CW-Radar), Deutsche Radar-Verl.-Ges., Garmisch, F.R.Germany, 216 p.

(81) Wheeler, G.J.: 1967. Radar Fundamentals. Prentice Hall, Englewood Cliffs (N.J.), 105 p.

(82) Woodward, P.M.: 1964. Probability and Information Theory with Application to Radar. Pergamon Press, London, p. 81–125.

(83) Wosnik, J.: 1965. Mikrowellentechnik und Antennen. Vieweg, Braunschweig (NTF 23), 312 p.

(84) Zemanian, A.H.: 1965. Distribution Theory and Transform Analysis. McGraw-Hill, New York, 371 p.

(85) Müller, H.: 1967. Astronomische Begriffe. Bibliograph. Inst., Mannheim, Bd. 57/57a, p. 90–95.

(86) Hovanessian, S.A.: 1980. Introduction to synthetic array and imaging radars. Artechn House, Dedham, Mass. USA, 156 p.

Papers in Periodicals (after appearance in the text)

(87) Ko, H.C.: 1964. Radio-Telescope Antenna Parameters. IEEE Transaction A.P., Vol. A.P., Dec. 12, p. 891–898.

(88) Ko, H.C.: 1964. Antenna Temperature and the Temperature of Electromagnetic Radiation. IEEE Trans., Vol. A.P., Jan. 12, p. 126–127.

(89) Bracewell, R.N., Roberts, J.A.: 1954. Aerial smoothing in radioastronomy. Australian J. Phys., Vol. 7, p. 615–640.

(90) MacPhie, R.: 1966, May. The Compound Intensity Interferometer. IEEE Trans., Vol. A.P., May 14, p. 369–374.

(91) Vitkevich, V.V.: 1953. Wide band radio interferometer. Dokl. Akad. Nauk., SSR, 91, 1301 p.

(92) Culshaw, W.: 1950. The Michelson Interferometer at mm Wavelength. Proc. Phys. Soc. 63-B, p. 393–460.

(93) Ko, H.C.: 1961, September. On the determination of the Disk Temperature and the Flux Density of a Radio Source Using High-Gain Antennas. IRE Trans., Vol. A.P.-9, p. 500–501.

(94) Swenson, G.W., Mathur, N.C.: 1968. The Interferometer in Radio Astronomy. Proc. IEEE, 56, Dec., p. 2114–2130.

(95) McCready, L.L., Pawsey, J.L., Payne-Scott, R.: 1947. Solar radio radiation and its relation to sunspots. Proc. Roy. Soc. A., 190, p. 357–360.

(96) Ryle, M., Vonberg, D.D.: 1946, September. Solar Radiation at 175 MHz. Nature, 158, p. 339–340.

(97) Ryle, M.: 1952. A New Interferometer and its Application to the Observation of weak Stars. Proceeding of the Royal Society, A, 211, p. 351–375.

(98) Peckham, R.J.: 1973. A Phase Correction Technique for Radio Link Interferometry. Mont. Not. Roy. Astr. Soc., No. 165, p. 25–38.

(99) Tseitlin, N.M.: 1970. Aperture Synthesis in Radio Astronomy (Review). Radio Energ. Electron. Physics, Vol. 15, No. 3, p. 369–387.

(100) Ryle, M., Neville, A.C.: 1963. A radio survey of the North Polar Region with a 4.5' of arc pencil beam System. Monthl. Notes. Roy. Astr. Soc., 125, p. 39–56.

(101) Sheridan, K.V.: 1963. Techniques for the Investigation of Solar Radio Bursts at Meter Wavelenghts, Proc. IRE Austral., Febr., 24, p. 174–184.

(102) Little, A.G., Payne-Scott, R.: 1951. The Position and Movement on the Solar Disk of Sources of Radiation at a Frequency of 97 MHz. Austral. J. Sci. Res. A, 4, Dec., p. 489–507.

(103) Christiansen, W.N., Warburton, J.A.: 1953. The Distribution of the Radio Brightness over the Solar Disk at a Wavelength of 21 cm. Austral. J. Phys., 6, p. 190–202.

(104) Christiansen, W.N., Warburton, J.A.: 1955, December. The Distribution of Radio Brightness over the Solar Disk at a Wavelength of 21 cm. Austr. J. Phys., 1, Vol. 8, p. 474–486.

(105) Christiansen, W.N., Labrum, N.R., McAlister, K.R., Mathewson, D.S.: 1961. The Cross-Grating Interferometer: a new High-Resolution Radio Telescope, Proc. IEE, Vol. 108, Pt.B, No. p. 748–758.

(106) Christiansen, W.N., Labrum, N.R., McAlister, K.R., Mathewson, D.S.: 1961, January. The Cross-Grating Interferometer: a new High-Resolution Radio Telescope. Proc. IEE, Pt.B, Vol. 108, p. 48–59.

(107) Mills, B.Y., Little, A.G.: 1953. A high resolution aerial system of a new type. Austral. J. Phys., 6, p. 272–280.

(108) Mills, B.Y.: 1963. Crosstype radio telescopes. Proc. IRE (Australia), Vol. 24, p. 132–139.

(109) Hidey, A.G.: 1963. Phase determination of coherence functions by the intensity interferometer. E.C. Jordan, Ed., Electromagnetic Waves, Vol. 6, Part 2. Electromagnetic Theory and Antennas. Pergamon Press, New York, p. 801–810.

(110) Christiansen, W.N., Mathewson, D.S.: 1958. Scanning the Sun with a Highly Directional Array. Proc. IRE (Australia), Vol. 46, p. 127–140.

(111) Christiansen, W.N.: 1972. The Fleurs Synthesis Radiotelescope. The Australian Physicist, Dec., p. 179–181.

(112) Christiansen, W.N.: 1976. Extensions to the Fleurs Synthesis Telescope. Proc. Astron. Society. Australia, Sept., p. 35.

(113) Christiansen, W.N.: 1973. The Fleurs Synthesis Telescope. Proc. IREE, Austral., Vol. 34, No. 8, Sept., p. 302–310.

(114) Bracewell, R.N., Swarup, G.: 1961. The Stanford Microwave Spectroheliograph Directional Array. IRE Trans., AP, 9, Jan., p. 22–30.

(115) Blum, E.: 1959. Sensibilité des radio telescopes et récepteurs à correlation. Ann. Astrophys., 22, p. 140–163.

(116) Hanbury-Brown, R., Twiss, R.Q.: 1954. A new type of interferometer for use in radioastronomy. Phil. Mag., Vol. 45, July, p. 663–682.

(117) Hanbury-Brown, R.: 1966. The stellar interferometer at Narrabri. Australia, Philips Technical Review, Vol. 27, No. 6, p. 141–159.

(118) Reed, I.A.: 1962. On a Moment Theorem for Complex Gaussian Processes. IRE Trans., Vol. IT-8, April, p. 194–195.

(119) Covington, A.E., Broten, N.W.: 1957. An Interferometer for Radioastronomy with a single-lobed Radiation Pattern. IRE-Trans., AP-5, p. 247–255.

(120) Clark, B.G., Weimer, R., Weinreb, S.: 1972. The Mark II VLBI System, Principles and Operating Procedures. NRAO EL. Div. Intern. Rep. No. 118, Green Bank, West Virginia, USA, April, 117 p.

(121) Kellermann, K.I.: 1972. Radio Galaxies, Quasars and Cosmology. The Astronomical Journal, Vol. 77, No. 7, Sept., p. 531–542.

(122) Klemperer, W.K.: 1972. Long-Baseline Radio Interferometry with Independent Frequency Standards. IEEE-Proc., 60, May, p. 602–609.

(123) Palmer, H.P., Rowson, B., Anderson, B., Donaldson, W., Miley, G.K., Gent, H., Algie, R.L., Slee, O.B., Crowther, H.H.: 1967. Radio Diameter Measurements with Interferometer Baselines of one Million and two Million Wavelengths. Nature, 213, Feb., p. 789–790.

(124) Broten, N.W., Legg, T.H., Locke, J.L., McLeish, C.W., Richards, R.W., Chisholm, R.M., Gush, H.P., Yen, J.L., Galt, J.A.: 1967. Long Baseline Interferometry: a new technique. Science, Vol. 156, June, p. 1592–1593.

(125) Broten, N.W., Legg, T.H., Locke, J.L., McLeish, C.W., Richards, R.S., Chisholm, R.M., Gush, H.P., Yen, J.L., Galt, J.A.: 1967. Observations of Quasars using Interferometer Baselines up to 3074 km. Nature, Vol. 215, p. 38.

(126) Broten, N.W., Clarke, R.W., Legg, T.H., Locke, J.L., McLeish, C.W., Richards, R.S., Yen, J.L., Chisholm, R.M., Galt, J.A.: 1967. Diameters of some Quasars at a Wavelength of 66.9 cm. Science, Vol. 216, p. 44.

(127) Broten, N.W., Clarke, R.W., Legg, T.H., Locke, J.L., Chisholm, R.M., Yen, J.L., Galt, J.A.: 1969. Long Baseline Interferometer Observations at 408 and 448 MHz. Monthl. Notices, Roy. Astr. Soc., No. 146, p. 313–327.

(128) Galt, J.A., Broten, N.W., Legg, T.H., Locke, J.L., Yen, J.L.: 1970. Long Baseline Interferometer Measurements of CP 0329. Nature, Vol. 225, No. 5232, Feb., p. 530–531.

(129) Bare, C., Clark, B.G., Kellermann, K.I., Cohen, M.H., Jauncey, D.L.: 1967. Interferometer experiment with independent local oscillators. Science, Vol. 157, July, p. 189–191.

(130) Van-Vleck, J.H., Middleton, D.: 1966. The Spectrum of Clippend Noise. Proc. IEEE, 54, Jan., p. 2–19.

(131) Basart, J.P., Clark, B.G., Kramer, J.S.: 1968. A Phase Stable Interferometer of 100000 Wavelength Baseline. Publ. of Astron. Soc. Pacific, Vol. 80, June, p. 273–280.

(132) Basart, J.P., Miley, G.K., Clark, B.G.: 1970. Phase Measurements with an Interferometer Baseline of 11.3 km. IEEE Trans., AP-18, May, p. 375–379.

(133) Cohen, M.H., Jauncey, D.L., Kellermann, K.I., Clark, B.G.: 1968. Radio Interferometry at one-thousandth second of arc. Science, Vol. 162, Oct., p. 88–94.

(134) Cannon, W.H.: 1978. The Classical Analysis of the Response of a Long Baseline Radio Interferometer. Geophys. J. Roy. Astr. Soc., 53, p. 503–530.

(135) Counselman III, C.C.: 1976. Radio Astrometry. Ann. Rev. of Astronomy and Astrophysics, Vol. 14, p. 197–214.

(136) Brosche, P.: 1977. On the Connection between Radio Positions and the FK4. The Astron. Journal, Vol. 82, No. 4, p. 296–301.

(137) MacPhie, R.H.: 1964. 1. On the Mapping by a Cross-Correlation Antenna System of Partially Coherent Radio Sources. IEEE-Trans., AP-12, Jan., p. 118–124.

(138) Utukuri, R.R.N., MacPhie, R.H.: 1967. Coincident Arrays for the Direct Measurement of the Principal Solution in Radio Astronomy. IEEE-Trans., AP-15, Jan., p. 49–59.

(139) MacPhie, R.H.: 1972. The Quasi-Linear Intensity Interferometer. IEEE-Trans., AP-20, No. 6, Nov., p. 755–763.

(140) MacPhie, R.H., Okongwu, E.H.: 1975. Spherical Harmonics and Earth-Rotation Synthesis in Radio Astronomy. IEEE-Trans., AP-23, No. 3, May, p. 386–391.

(141) Swenson, G.W.Jr.: 1969. Synthetic-Aperture Radio Telescopes. Ann. Rev. Astron. Astrophysics, Vol. 7, p. 353–374.

(142) Wink, J.: 1973. Radioastronomische Untersuchungen der Strukturen ausgewählter HII-Regionen mit Hilfe der Apertursynthese-Technik, Dissertation, Fachbereich Physik, University of Muenster/Germany, p. 1–61.

(143) Goldstein, S.J.: 1970. Angular Resolution of Long Baseline Interferometers. NRAO-VLBI-Symp., Charlottsville, VA, USA, April.

(144) Swenson, G.W., Kellermann, K.I.: 1975. An Intercontinental Array - A Next-Generation Radiotelescope. Science, Vol. 188, June 27, p. 1263–1268.

(145) Kellermann, K.I.: 1974. Intercontinental Radio Astronomy. Scientific American, Vol. 226, Feb., p. 72–83.

(146) Rogers, A.E.E.: 1970. Very long Baseline Interferometry with Large Effective Bandwidth for Phase-Delay Measurements. Radio Sci., Vol. 5, Oct., p. 1239–1248.

(147) Shapiro, I.I., Robertson, D.S., Knight, C.A., Counselman, C.C., Rogers, A.E.E., Hinteregger, H.F., Lippincott, S., Whitney, A.R., Clark, T.A., Niell, A.E., Spitzmesser, D.J.: 1974. Transcontinental Baselines and the Rotation of the Earth measured by Radio Interferometry. 1974, Science, Vol. 189, p. 920–922, 191, 451.

(148) Ryle, M., Hewish, A.: 1960. A Synthesis of a large Radiotelescope. 1960, Month. Notes Royal Astr. Soc., 120, p. 220–230.

(149) Ryle, M.: 1962, May 12. The New Cambridge Radio Telescope. Nature, 1962, No. 194, p. 517–518.

(150) Ryle, M., Hewish, A., Shakeshaft, J.R.: 1959. The Synthesis of Large Radio Telescopes by the use of Radio-Interferometers. IRE-Transactions AP-7, Dec., p. 120–124.

(151) Hoegboem, J.A.: 1974. Aperture Synthesis with a Non-Regular Distribution of Interferometer Baselines. Astron. Astrophys. Suppl. 1974, 15, p. 417–426.

(152) Thiel, M.A.F.: 1979. The Gausz taper or "zooming" method. Priv. Communic., MPIfR, Bonn, FRG, May.

(153) Jennison, R.C.: 1958. A Phase Sensitive Interferometer Technique for the measurement of the Fourier Transforms of Spatial Brightness Distribution of Small Angular Extent. Monthl. Not. Roy. Astr. Soc., March, p. 276–284.

(154) Rogers, A.E.E., Hinteregger, H.F., Whitney, A.R., Counselman III, C.C., Shapiro, I.I., Wittels, J.J., Klemperer, W.K., Warnock, W.W., Clark, T.A., Hutton, I.K., Marandino, G.E., Roennaeng, B.O., Rydbeck, E.H., Niell, A.E.: 1974. The Structure of Radio Sources 3 C 273 B and 3 C 84 deduced from the "Closure" Phases and Visibility Amplitudes Observed with three-element Interferometers. Ap. Journ. Vol. 193, p. 293–301.

(155) Preuss, E.: 1976, 1977. Hochaufloesende Radiointerferometrie (VLBI), Sterne u. Weltraum, Vol. 15, p. 246–250, 392–396, Vol. 16, p. 86–90.

(156) Schwartz, R., Witzel, A.: 1977. Radioteleskope im weltweiten Verbund, Blick ins Herz der Galaxien. Bild der Wissenschaft, 11, p. 158–176.

(157) Wilkinson, P.N., Readhead, A.C.S., Purcell, G.H., Anderson, B.: 1977, October. The Radio Structure of 3 C 147 Determined by Multi-Element Very-Long-Baseline Interferometry. Nature, p. 764–770.

(158) Wohlleben, R., Wielebinski, R., Mattes, H.: 1971. Feeds for the 100 m Effelsberg Telescope. Proceedings of the 2. European Microwave Conference. Stockholm, August, p. B5/5:1–B5/5:4.

(159) Wohlleben, R., Mattes, H., Lochner, O.: 1972. A Simple Small Primary Feed for Large Opening Angles an High Aperture Efficiency. Electronics Letters, Oct. 20, p. 181–183. (LOVE, A. Electromagnetic Horn Antennas, IEEE-Pr., N.Y. 1975, 91 p.)

(160) Wohlleben, R., Mattes, H.: 1971, Nov. 15. Primaerfokus-Erreger fuer das 100 m Radioteleskop. Kleinheubacher Berichte No. 15, 1971, p. 413–419.

(161) Koch, G.F., Lochner, O., Scheffer, H., Wohlleben, R.: 1975, February. Untersuchungen am Radioteleskop Effelsberg mit neuen Erregern. Nachrichten. Zeits., No. 28, H. 2, p. 42–46.

(162) Wohlleben, R., Noll, T.: 1975. Two primary feeds for dual beam operation in reflector antennas. IEEE-AP-S-Int. Sympos. Digest, Urbana/Ill. USA, 1.-5.6, p. 387–390.

(163) Wohlleben, R., Noll, T.: 1977. Primaerfokus-Erreger mit geringem Rueckstreuquerschnitt fuer Parabolreflektoren. Nachr. Techn. Fachber., NTF 57, Berlin, VDE-Verlag, p. 85–90.

(164) Wohlleben, R.: 1974. Quasi-Hybridhorn mit wenigen Fallen. Nachrichtentechn. Zeits., Bd. 27, May, H. 5, p. 168–170.

(165) Wohlleben, R., Mattes, H.: 1972. Doppeldipol-Erreger fuer den Primaerfokus des Radioteleskops Effelsberg. Nachr. Techn. Fachber., NTF 45, VDE-Verlag, Berlin, p. 95–97.

(166) Wohlleben, R., Baack, C.: 1978. Das V-Dipol-Paar als Erreger fuer tiefe Parabolantennen. Frequenz, H. 7, July, p. 200–204.

(167) Wohlleben, R., Koehler, J.A.: 1983. A New Blocking Condition including Subreflector and Quadrupol Shadow for the Design of Dual Reflector Systems. IEEE, Trans., AP-U, March, p. 342–346.

(168) Wohlleben, R.: 1974. Erreger fuer den Primaerfokus des Radioteleskops Effelsberg. MPIfR-El. Abt., Techn. Ber. No. 8, Sept., p. 1–65.

(169) Wohlleben, R.: 1987. Primary feeds for the 100-m Effelsberg Radiotelescope. Proc. ESA-ESTEC Workshop "Primary feeds and RF sensing system" Noordwijk/NL, 10./11.87, p. 25–28.

(170) Rusch, W.V.T., Wohlleben, R.: 1981. Relative Influence of Subreflector and Main reflector Surface Errors an Overall Tolerance Loss of a Dual-Reflector Antenna. IEEE Trans., AP-29, Aug., p. 784–785.

(171) Wohlleben, R., Mutschlechner, J.: Sliding hybrid mode feed for variable illumination of reflector antennas. Proc. MIOP, May 1987, Vol. 3, p. 34–35.

(172) Cohen, M.H., Kellermann, K.I., Shaffer, D.B., Linfield, R.P., Moffet, A.T., Romney, J.D., Seielstad, G.A., Pauliny-Toth, I.I.K., Preuss, E., Witzel, A., Schilizzi, R.T., Geldzahler, B.: 1977. Radio Sources with Superluminal Velocities. Review Article, Nature, Vol. 268, Aug. 4, p. 405–409.

(173) Kellermann, K.I., Clark, B.G., Jauncey, D.L., Cohen, M.H., Shaffer, D.B., Moffet, A.T., Gulkis, S.: 1970. High-Resolution Observations of Compact Radio Sources at 13 centimeters. Ap. Jour., Vol. 161, p. 803–809.

(174) Swenson, G.W., Mathur, N.C.: 1969. On the space-frequency Equivalence of a Correlator Interferometer. NBS, Rad. Science, Vol. 4, Jan., p. 69–71.

(175) Pauliny-Toth, I.I.K, Preuss, E., Witzel, A., Kellermann, K.I., Shaffer, D.B., Purcell, G.H., Grove, G.W., Jones, D.L., Cohen, M.H., Moffet, A.T., Romney, J., Schilizzi, R.T., Rinehart, R.: 1976. High resolution observations of NGC 1275 with a four-element intercontinental radio interferometer. Nature, No. 259, Jan. 1, p. 17–20.

(176) Kellermann, K.I., Shaffer, D.B., Pauliny-Toth, I.I.K., Preuss, E., Witzel, A.: 1976. Observations of a Radio Source in the Nucleus of M 81 with Dimensions Less than 1300 Astronomical Units. The Astrophys. Journ., No. 210, Dec. 15, p. L121–L122.

(177) Pauliny-Toth, I.I.K., Preuss, E., Witzel, A.: 1975. Observations of Compact Radio Sources by Means of Very-Long-Baseline Interferometry. Kleinheubacher Berichte, Vol. 19, p. 327–331.

(178) Clark, T.A., Counselman III, C.C., Ford, P.G., Hanson, L.B., Hinteregger, H.F., Klepczinski, W., Knight, C.A., Robertson, D.S., Rogers, A.E.E., Ryan, J., Shapiro, I.I., Whitney, A.R.: 1977. Precise Clock Synchronization via very-Long-Baseline Interferometry. 8. Precise Time T. Int. Meeting, Goddard Space Flight Center, Dec., p. 349–359.

(179) Counselman III, C.C., Hinteregger, H.F., Shapiro, I.I.: 1972. 10. Astronomical applications of differential interferometry. Science, Vol. 178, Nov., p. 607–608.

(180) Counselman III, C.C., Hinteregger, H.F., King, R.W., Shapiro, I.I.: 1973. Determination of lunar baselines and libration by differential VLBI observations of ASESPS. The Moon, Vol. 8, p. 484–489.

(181) Whitney, A.R., Rogers, A.E.E., Hinteregger, H.F., Knight, C.A., Levine, J.I., Lippin-Cott, S., Clark, T.A., Shapiro, I.I., Robertson, D.S.: 1976. A Very-Long-Baseline interferometer system for geodetic applications. Rad. Science, Vol. 11, No. 5, May, p. 421–432.

(182) Hinteregger, H.F., Shapiro, I.I., Robertson, D.S., Knight, C.A., Ergas, R.A., Whitney, A.R., Rogers, A.E.E., Moran, J.M., Clark, T.A., Burke, B.F.: 1972. Precision geodesy via radio astronomy. Science, Vol. 178, p. 396–398.

(183) Kellermann, K.I., Jauncey, D.L., Cohen, M.H., Shaffer, B.B., Clark, B.G., Broderick, J., Rönnäng, B., Rydbeck, O.E.H., Matvenyenko, L., Moiseyeo, I., Vitkevitch, V.V., Cooper, P.F.C., Batchelor, R.: 1971, October. High-resolution observations of compact radio sources at 6 and 18 centimeters. The Astrophysical Journal, 169, p. 1–24.

(184) Shapiro, L.D., Fisher, D.O.: 1970. Using Loran-C transmissions for long baseline synchronisation. Radio Science, 5, p. 1233–1250.

(185) Cronyn, W.M.: 1972. Interferometer Visibility Scintillation. Astrophys. Journal, Vol. 174, May 15, p. 181–200.

(186) Galt, J.A., Broten, H.W., Legg, T.H.: 1975. Simultaneous Interferometric and Spectrometric Observations of Pulsar 0329+54. The Astron. Journ., Vol. 80, No. 4, April, p. 311–317.

(187) Wood, L.E., Thompson, M.C.: 1970. Oscillator synchronization via satellite. Radio Science, 5, Oct., p. 1249–1252.

(188) Moran, J.M.: 1973. Spectral-Line Analysis of Very-Long-Baseline Interferometric Data. IEEE-Proc., 61, Sept., p. 1236–1242.

(189) Blythe, M.: 1957. A new Type of Pencil Beam Aerial for Radio Astronomy. Monthly Notes of the Royal Astronomy Society, No. 117, p. 644–651.

(190) Ryle, M.: 1957. The Mullard Radio Astronomy Observatory, Cambridge. Nature, No. 180, July, p. 110–112.

(191) Baldwin, J.E., Field, C., Warner, P.J., Wright, M.C.H.: 1971. Mapping of Neutral Hydrogen. Mon. Not. Roy. Astr. Soc., No. 154, p. 445–454.

(192) Ryle, M.: 1967. The Cambridge 1 mile radio Telescope. Electronics and Power, 13, June, p. 208–211.

(193) Hinder, R., Ryle, M.: 1971. Atmospheric limitations to the angular resolution of aperture synthesis radio telescopes. Mon. Not. Roy. Astr. Soc., No. 154, p. 229–253.

(194) Elsmore, B., Ryle, M.: 1976. Further Astrometric Observations with the 5-km Radio Telescope. Mon. Not. Roy. Astr. Soc., 174, p. 411–425.

(195) Ryle, M.: 1974. Radio and Optical Studies of 4C11.50. Mon Not. Roy. Astr. Soc., 168, p. 1p–6p.

(196) Ryle, M.: 1972. The 5-km Radio Telescope at Cambridge. Nature, Vol. 239, Oct., p. 435–438.

(197) Ryle, M.: 1961. Radio Astronomy and Cosmology. Nature, No. 190. 3, June, p. 852–854.

(198) Ryle, M., Elsmore, B., Neville, A.C.: 1965. High Resolution Observations of the Radio Sources in Cygnus and Cassiopeia. Nature, No. 205, p. 1259–1262.

(199) Laroye, J.F.: 1979. Phase Comparison via Satellite Link. MS-Thesis, Appl. Science, Dept. El. Engrg. Univ. of Toronto. Jan., 81 p.

(200) Ryle, M.: 1975. Radio Telescopes of Large Resolving Power. Science, 188, June 13, p. 1071–1079.

(201) Weiler, K.W., Raimond, E.: 1976. Aperture Synthesis Observations of Circular Polarization: I. Methods and their Application to the Observation of Unresolved Sources at 1.4 GHz. Astr. Astrophys. 52, p. 397–402.

(202) Conway, R.G., Kronberg, P.P.: 1969. Interferometric Measurement of Polarization Distributions in Radio Sources. Mont. Not. Roy. Astr. Soc., 142, p. 11–32.

(203) Argue, A.N., Ekers, R.D., Fanaroff, B.L., Hazard, C., Ryle, M., Shakeshaft, J.R., Stockton, A., Webster, A.S.: 1974. Radio and Optical Studies of 4011.50. Mon. Not. Roy. Astr. Soc., 168, Short Comm., p. 1p–6p.

(204) Mitton, S., Ryle, M.: 1969. High resolution observation of Cygnus A at 2.7 and 5 GHz. Mon. Not. Roy. Astr. Soc., 146, p. 221–233.

(205) Hall, D.N., Learner, R.C.M., Barr, L.D.: 1977. Next Generation Telescope. Kitt Peak National Obs. Report No. 1, Tucson Arizona/USA, Feb., p. 1–31.

(206) Spizzichino, A., Delcourt, J., Giraud, A., Revah, I.: 1965. A new type of continuous wave radar for the observation of meteor trails. Proc. IEEE, Vol. 53, Aug., p. 1084–1086.

(207) Ulbricht, G.: 1968. Funkortung in der Raumfahrt. Lecture Notes, Tech. Univ. Berlin, 180 p.

(208) Wheeler, M.S., Hackers, P.S.: 1962. Interferometer is Disposed for Gemini Radar. Electronics, Nov. 30, p. 106–108.

(209) Bachmann, W., Hissen, H.: 1968. Fortschritte der Radartechnik durch Modulationsverfahren. In Bücherei der Ortung und Navigation, Nr. 101, Deutsche Gesellschaft für Ortung und Navigation, Düsseldorf.

(210) Hissen, H.: 1969. Zur Dimensionierung von Radarsignalen. A. Elek. Übertr., 23, H. 7, p. 381–393.

(211) Bachmann, W.: 1969. Modulation von Radar- und Sonarimpulsen durch binäre Frequenzumtastung. Arch. Elektr. Übertrag., 23, H. 5, p. 251–257.

(212) Bachmann, W.: 1968. Beitrag zum Verständnis der Ambiguityfunktion phasengetasteter Impulse. Nachrichtentech. Zeitschrift (NTZ), 5, p. 278–282.

(213) Hissen, H.: 1970. Ortung mit Multifrequenzsignalen. Diss. Fak. f. Elektrotechnik, RWTH Aachen, Germany, July, 130 p.

(214) Pahls, J.: 1970. Nebenmaxima der Ambiguityfunktion binär phasengetasteter Multifrequenz-Radarsignale. Arch. Elektr. Übertrag., 23, p. 164–168.

(215) Pahls, J.: 1970. Impulskompression zur Radarverstärkung. Analysis und Verarbeitung binär angetasteter Signale. Diss. Fak. f. Elektrotechnik, RWTH Aachen, FRG, July, 107 p.

(216) Allen, J.L.: 1962. Array radars - survey of their potential and the limitations. Microwave Journal, Vol. 5, May, p. 67–79.

(217) Difranco, I.V., Rubin, W.I.: 1963. A high resolution interferometer radar with low angle ambiguity. IRE Trans., Vol. A.P.-11, March, p. 197–198.

(218) Brookner, E.: 1964. Multidimensional Ambiguity Functions of Linear Interferometer Antenna Arrays. IEEE Trans., A.P.-12, Sept., p. 551–561.

(219) Kosel, G.: 1969. Radarempfänger für ein Rundsichtgerät mit Minimumpeilung. Diss., Fakultät f. Elektrotechnik, RWTH Aachen, Germany, June, 127 p.

(220) Achilles, D.: 1969. Impulskompression mit Hermiteschen Funktionen. Nachrichtentechn. Zeitschrift, H. 5, p. 276–280.

(221) Marchie, R.H.: 1962. Optimum Cross Correlation Radar System. Proc. IRE, Vol. 50, p. 2508–2509.

(222) MacPhie, R.H.: 1966. A Phase-switched Radar System Giving Improved Control of Directional Pattern. Radio Electron. Engr. G.B., Vol. 30, Feb., p. 81–92.

(223) Mayo, B.P., Adams, W.S.: 1963. Synthetic Aperture Antenna Investigation. RADC, ARDC, USAF, Griffis Airforce Base, Rome, N.Y., Final Report Contract AF 30 602–2323.

(224) Paul, R.H.: 1968. Refraction Error in Measurement of Angle with the Radio Interferometer. IEEE Trans., Vol. AES-4, Jan. 1, No. 1, p. 52–57.

(225) Paul, R.H.: 1969. A Comparison of Radar and Radio Interferometer Refraction Errors. IEEE Trans., Vol. AES-5, No. 2, March, p. 346–350.

(226) Callender, D.W.: 1967. Microwave Interferometer Calibrates Aircraft Instrument Landing System. Rept. Aerospace. Defense and Marine Products Division, Westinghouse Elect. International Co., USA.

(227) Simmons, G.A.: 1957. A Theoretical Study of Errors in Radio Interferometer. Type Measurements Attributable to Inhomogenities in the Medium. IRE Trans., Vol. TRC-3, Dec., p. 2–5.

(228) Bean, B.R., Thayer, G.D.: 1959. CRPL exponentail reference atmosphere. Central Radio Lab. Nat. Bur. Standards. Boulder, Col., Monograph No. 4.

(229) Bean, B.R., Thayer, G.D.: 1959. Models of Atmospheric Radio Refractive Index. Proc. IRE, Vol. 47, p. 740–755.

(230) Barton, D.K.: 1965. Multipath Error in a Vertical Interferometer. Proc. IEEE, Vol. 53, No. 5, May, p. 543–544.

(231) Baars, J.W.M.: 1967. Meteorological Influences on Radio Interferometer Phase Fluctuations. IEEE Trans., Ap-15, July, p. 582–584.

(232) Baars, J.W.M.: 1970. Dual Beam Parabolic Antennae in Radioastronomy. Diss. Univ. Leiden Wolter-Noordhoff, Groningen, The Netherlands, 128 p.

(233) Cohen, W., Steinmetz, C.M.: 1959. Amplitude and phase sensing monopulse system parameter. Part. II, Microwave Journal, Vol. 2, Oct., p. 27–33.

(234) D'Antonio, R.A., Gaffney, J.E.: 1968. Pseudo-Noise-Interferometer. IEEE Trans., Vol. AES-4, Sept., No. 5, p. 728–732.

(235) Simmons, G.J.: 1959. A Study of the Error Involved in the Near Use of a Radio Interferometer. Techn. Memo., Scandia Corp. Albuquerque, N.M., April, USA.

(236) Sparagna, J.J.: 1967, July. DF Vectoring: Applications to Stationary Synchronous Satellites. IEEE Trans., Vol. AES-3, No. 4, p. 697–704.

(237) Rumsey, V.H., Deschamps, G.A., Kales, M.L., Bohnert, J.L.: 1951. Techniques for Handling Elliptically Polarized Waves with Special Reference to Antennas. Proc. IRE, Vol. 39, May, p. 533–556.

(238) Wheeler, M.S.: 1964. Phase Characteristics of Spiral Antennas for Interferometer Applications. IEEE Internat. Convention Record, Part 2, p. 143–147.

(239) Carl, J.S., Guccione, S.A.: 1966. Gemini Rendezvous Radar Error. The Microwave Journal, June, p. 75–78.

(240) Genzel, R.: 1978. Beobachtung von H_2O-Masern in Gebieten von OB-Sternentstehung. Diss. Math.-Naturw. Fakultät, Univ. Bonn, F.R.Germany, Nov., p. 101–133.

(241) Pauliny-Toth, I.I.K., Preuss, E., Witzel, A., Genzel, R., Kellermann, K.I., Shaffer, D.B., Matveyenko, L.I., Moiseev, I.G., Kogan, L.R., Kostenko, V.I., Ephanov, V.A., Roennaeng, B.: 1978. High-resolution Observations of Compact Radio Sources at 1.35 cm Wavelength. Astronomiceskij Zhurnal, Vol. 4, H. 2, p. 64–69, english translation in: Sovjet Astronomy, Journ. Letters, 4, 1, p. 32–35.

(242) Preuss, E., Kellermann, K.I., Pauliny-Toth, I.I.K., Witzel, A., Shaffer, D.B.: 1979. Structural Changes in the Nucleus of NGC 1275 at 2.8 cm Wavelength. Astronom. and Astrophys., p. 268–273.

(243) Westerbork Rad. Obs., Annual Report: 1977. Westerbork Nov. 1977, p. 124–140.

(244) N.R.A.O., A Very Large Array Radio Telescope. Vol. I, II, 1967, Jan., Green Bank, USA.

(245) Hachenberg, O.: 1968. Betrachtungen zum Bau grosser Radioteleskope. Westdt. Verlag, Leverkusen, F.R.G., Heft 177, Arb. Geb. f. Forschg. N.W., 44 p.

(246) Feix, G.: 1968. Radioastronomische Interferometrie mit extrem langer Basislänge. Nachrichtentechn. Zeitsch., H. 5, p. 263–266.

(247) Moffet, A.T., Gubbay, J.S., Robertson, D.S., Legg, A.J.: 1972. External Galaxies and Quasi-Stellar Objects. ed. D.S. Evans, Reidel/Dordrecht (Netherlands) p. 228–235.

(248) Hooghout, B.G.: 1966. A Synthesis Radio Telescope at Westerbork. Holland. "Design and Construction of Large Steerable Aerials". London, IEE, Conf. Publ. No. 21, June, p. 329–334.

(249) Hooghout, B.G.: 1964. The Benelux Cross Antenna Project. Ann. N.Y. Acad. Sci., 116, 1, p. 13–24.

(250) Tanaka, H., Kakinuma, T.: 1965. Improvement of the High-Resolution Interferometer at 9.4 GHz. Proc. Research Inst. of Atmospherics, Nagoya Univ., Toyokawa/Japan, Vol. 12, p. 27–34.

(251) Schilizzi, R.T.: 1978. Radio Astrometry Using VLBI Techniques. Proc. Coll. Europ. Satellite Astrometry, Padova/Italy, June, 5-7, p. 1–13.

(252) Yen, J.L., Kellermann, K.I., Rayher, B., Broten, N.W., Fort, D.N., Knowles, S.H., Waltman, W.B., Swenson, G.W.: 1977. Real-Time Very-Long-Baseline Interferometry Based on the Use of a Communication Satellite. Science, Vol. 1, 98, Oct. 21, p. 289–291.

(253) Witzel, A., Schalinski, C.J., Johnston, K.J., Biermann, P.L., Krichbaum, T.P., Hummel, C.A., Eckart, A.: 1988. The occurrence of bulk relativistic motion in compact radio sources. Astron. Astrophys. 206, 245–252.

(254) Clark, B.G.: 1968. Radio Interferometers of Intermediate Type. IEEE-Trans., AP-16, 2, p. 143–144.

(255) Ferrari, P., Pacholczyk, A.G.: 1983. Astrophysical Jets. Reidel Publ., Dordrecht, Netherlands, 280 p.

(256) Welichajew, L.: 1981. Signal Processing in a Phase Coherent Interferometer. IRAM, Grenoble/France, Work. Rep. No. 67, 16 p.

(257) Welichajew, L.: 1985. An Overview of the IRAM Interferometer. Proc. of Int. Symp. on MM and Submm. Wave Radio Astronomy, URSI-Granada, p. 89–97.

(258) Moran, J.M.: 1985. The SAO Submillimeter Telescope Array Project. Proc. Of. Int. Symp. on MM and Submm-Wave Radio Astronomy, URSI-Granada, p. 85–88.

(259) Ishigoro, M.: 1981. The Nobeyama mm-wave 5-element synthesis telescope. Nobeyama Rad. Obs., Rep. No. 1 and 7.

(260) Masson, C.R., Berge, G.L., Claussen, M.J., Heiligman, G.M., Leighton, R.B., Lo, K.Y., Moffer, A.T., Phillips, T.G., Sargent, A.I., Scott, S.L., Woody, D.P., Young, A.: 1985. The Caltech Millimeter Wave Interferometer. Proc. of Int. Symp. on MM and Submm. Wave Radio Astronomy, URSI-Granada, p. 65–74.

(261) Wild, J.P., Sheridan, K.V.: 1958. A Swept-Frequency Interferometer for the Study of High-Intensity Solar Radiation at Meter Wavelengths. Proc. IRE, Jan., p. 160–171.

(262) Stirner, E.: 1977/85. Antennen. Vol. 1-3, Huethig, Heidelberg 229, 215 and 203 p.

(263) Lovell, B.: 1974. Out of the Zenith. Oxford Univ. Press, London, pp. 9–44, 254.

(264) Hachenberg, O.: 1968. Studien zur Konstruktion des 100-m-Teleskops. Beitraege zur Radioastronomie, Duemmler, Bonn, Dez., 1, H. 2, 61 p.

(265) Goessl, H. (Ed.): 1976. Das Radiointerferometer der DFVLR (at 137 MHz). Forsch. Ber., Bu. Min. FT/Munich, Dec., 166 p.

(266) Nollet, M., Verfaillie, G., Goessl, H., Houmann, D.: 1974. Advanced VHF Interferometer Spacecraft Tracking System. Electrical Communication, ITT/Brussels, Vol. 49, No. 3, p. 204–217.

(267) Napier, P.J., Thompson, A.R., Ekers, R.D.: The Very Large Array: 1983. Design and Performance of a Modern Synthesis Radio Telescope. Proc. IEEE, 71, Nov., p. 1295–1320.

(268) Kellermann, K.I.: 1977. VLBI-Network Studies III (An Intercontinental Very Long Baseline Array, VLBA). Natl. Radio Astron. Observatory, Green Bank, W.VA. USA, May, 136 p.

(269) Swenson, G.W.Jr.: 1977. VLBI-Network Studies Vol. IV (On the Geometry of the VLBI Network, later: VLBA). Univ. of Ill., Urbana, Champaign/Ill., Grant: NASA, AST 76-24650, Dec., 93 p.

(270) Shapero, D. (Ed.): 1983. Multidiscipline use of the Very Long Baseline Array. Proc. of a Workshop of Board of Phys. and Astronomy of Natl. Res. Council, US-Natl. Academy Press, Washington, 202 p.

(271) Marcaide, J.M.: 1982. VLBI Studies of the Extragalactic Radio Sources 1038+528 A, B, Ph.-D. diss. MIT, August, 186 p.

(272) Napier, P.J.: 1982. A Revised Proposal for a Feed System for the VLBA Antenna (element). LVB Array Memo No. 59, Nov. 1, VLA- Soccorro/NM-USA, 6 p.

(273) Anderson, A.J.: 1979. Scandinavian Studies of Recent Crustal Movements and the Space Geodetic Baseline Network. Rep. No. 4, Geodetic Inst., Univ. Uppsala/Sweden, or: Proceed. "Radio Interferometry Techniques in Geodesy" Coates, R.J. (Ed.) Goddard Space Flight Ctr./USA, 1979, 12 p.

(274) Spoelstra, T.A.T.: 1983. The Influence of ionospheric refraction on radio astronomy interferometry. Astronomy and Astrophysics, 120, p. 313–321.

(275) Spoelstra, T.A.T.: 1985. Effects of amplitude and phase scintillations on decimeter wavelength observations at mid-latitudes. Astron. Astrophys., 148, p. 21–28.

(276) Spoelstra, T.A.T.: 1986. Correcting refraction in radio astronomy. Publ. Astron. Observ. Beograd/Yugoslavia, p. 1–43.

(277) Olthof, H. (Ed.): 1985. QUASAT a space VLBI satellite. Assessment Study, SCI-ESA-ESTEC, Noordwijk/NL, Nov., 95 p.

(278) Rogers, A.E., Cappallo, R.J., Hinteregger, H.F., Levine, J.I., Nesman, E.F., Webber, J.C., Whitney, A.R., Clark, T.A., Ma, C., Ryan, J., Corey, B.E., Counselman, C.C., Herring, T.A., Shapiro, I.I., Knight, C.A., Sheffer, D.B., Vandenberg, N.R., Lacasse, R., Maury, R., Rayhrer, B., Schupler, B.R., Pigg, J.C.: 1983. Very-Long Baseline Radio Interferometry: The Mark III System for Geodesy, Astrometry and Aperture Synthesis. Science, 219, Jan. 7, p. 51–54.

(279) Kellermann, K.I.: 1980. Radio Galaxies and Quasars. Annals of the N.Y. Academy of Sciences, 336, Feb. 15, p. 1–11.

(280) Kellermann, K.I., Pauliny-Toth, I.I.K.: 1981. Compact Radio Sources. Annales Rev. Astron. Astrophys., 19, p. 373–410.

(281) Porcas, R.W., Booth, R.S., Browne, I.W.A., Walsh, D., Wilkinson, P.N.: 1981. VLBI structures of the images of the double QSO-0957+561. Nature, Vol. 289, Feb. 26, p. 758–762.

(282) Morabito, D.D.: 1984. Submilliarcsecond VLBI Using Compact Close Pairs of Radio Sources: Error Analysis. Jet. Prop. Lab./Pasadena, TDA Progress Report No. 42-79, July, p. 1–16.

(283) Kellermann, K.I., Downes, A.J.B., Pauliny-Toth, I.I.K., Preuss, E., Shaffer, D.B., Witzel, A.: 1981. VLBI Observations of the Nucleus of the Radio-Galaxy Cygnus A. Astronomy and Astrophys., 97, p. L1–L4.

(284) Marathay, A.S.: 1976. Radiometry of Partially Coherent Fields. Proc. Ann. Meet. Opt. Soc. Am. Tucson, Oct. 18-22, paper: TuE 14.

(285) O'Neill, E.L.: 1963. Introduction to Statistical Optics. Addison-Wesley, Reading/Mass. USA, p. 80–85.

(286) Beran, M.J., Parrent, G.B.Jr.: 1974. Theory of Partial Coherence. Society of Photo-optical Instrumentation engineers, Prent. Hall, Englewood Cliffs/USA, p. 159–174.

(287) Troup, G.J.: 1967. Optical Coherence Theory-Recent Developments. Methuen, London, 76 p.

(288) Marcaide, J.M., Shapiro, I.I., Corey, B.E., Cotton, W.D., Gorenstein, M.V., Rogers, A.E.E., Romney, J.D., Schild, R., Baath, L., Bartel, N., Cohen, N.L., Clark, T.A., Preston, R.A., Ratner, M.I., Whitney, A.R.: 1984. On the quasars 1038+528 A and B. Astron. Astrophys., Nov., 41 p.

(289) Clark, T.A., Corey, B.E., Davis, J.L., Elgered, G., Herring, T.A., Hinteregger, H.F., Knight, C.A., Levine, J.I., Lundquist, G., Ma, C., Nesman, E.F., Phillips, R.B., Rogers, A.E., Roennaeng, B.O., Ryan, J.W., Schupler, B.R., Schaffer, D.B., Shapiro, I.I., Vandenberg, N.R., Webber, J.C., Whitney, A.R.: 1985. Precision Geodesy Using the Mark-III Very-Long-Baseline Interferometer System. IEEE Trans., GE-23, No. 4, July, p. 438–449.

(290) Bartel, N., Dhawan, V., Krichbaum, T., Graham, D.A., Pauliny-Toth, I.I.K., Rogers, A.E.E., Roennaeng, B.O., Spencer, J.H., Hirabayashi, H., Inoue, M., Lawrence, C.R., Shapiro, I.I., Burke, B.F., Marcaide, J.M., Johnston, K.J., Booth, R.S., Witzel, A., Morimoto, M., Readhead, A.C.S.: 1988. VBLI imaging with an angular resolution of 100 microarcseconds. Nature, 334, 14 July, p. 131–136.

(291) Parrent, G.B.: 1959. Studies in the theory of partial coherence. Optica Acta 6, p. 285–296.

(292) Thompson, A.R., Moran, J.M., Swenson, G.W.: 1986. Interferometer and Synthesis in Radio Astronomy. Wiley, New York, 435 p.

(293) Bracewell, R.: 1978. The Fourier Transform and its applications. McGraw-Hill, New York, 329 p.

(294) Clark, B.G.: 1973. The NRAO Tape Recorder Interferometer System. Proc. IEEE 61, p. 1242–1248.

(295) Hellwig, H.: 1979. Microwave Time and Frequency Standards. Radio Science 14, p.

(296) Fanti, R., Kellermann, K., Setti, G.: 1984. VLBI and Compact Radio Sources. Proc. of the IAU Symp. No. 110, Reidel, Dordrecht.

(297) Smith, D.H.: 1984. MERLIN: A Wizard of a Telescope, Sky and Telescope, January, p. 31–32.

(298) Schwab, F.R.: 1980. Adaptive calibration of radio interferometer data. Soc. of Photo-Optic. Inst. Eng. 231, p. 18–24.

(299) Cornwell, T.J., Wilkinson, P.N.: 1981. A new method for making maps with unstable radio interferometry. Mon. Not. Roy. Astron. Soc. 196, p. 1067–1086.

(300) Khachikian, E.Ye., Melnick, J., Fricke, K.J.: 1987. Observational Evidence of Activity in Galaxies. Proc. of the IAU Symp. No. 121, Reidel, Dordrecht.

(301) Schwarz, V.J.: 1979. The method 'clean'-use, misuse and variations. in C. van Schoeneveld: Image Formations from coherence Functions in Astronomy. Reidel, Dordrecht, p. 261–275.

(302) Narayan, R., Nityanda, R.: 1986. Maximum entropy image resolution in Astronomy. Ann. Rev. Astron. Astrophys. 24, p. 127– 170.

(303) Jaynes, E.T.: 1968. Prior Probabilities. IEEE Trans. Syst. and Sci. Cybern. SSC-4, p. 227–241.

(304) Jaynes, E.T.: 1982. The rationale of maximum entropy methods. Proc. IEEE 70, p. 939–952.

(305) Kikuchi, R., Soffer, B.H.: 1977. Maximum entropy image restauration. Journ. Opt. Soc. America 67-2, p. 1656–1665.

(306) Genzel, R., Reid, M.J., Moran, J.M., Downes, D.: 1981. Proper Motions and Distances of H_2O MASER Sources. I. The outflow in Orion-KL. Astrophys. J. 244, p. 884–902.

(307) Moran, J., Burke, B.F., Barrett, T., Rogers, A.E.E., Ball, J.A., Carter, J.C., Cudaback, D.D.: 1968. The Structure of the OH Source in W3. The Astr. Journ. 152, p. L97–L100.

(308) Genzel, R. et al.: 1978. Structure and Kinematics of H_2O Sources in Clusters of newly formed OB stars. Astron. Astrophys. 66, p. 13–29.

(309) Cohen, M.H., Shaffer, D.B.: 1971. Positions of Radio Sources from Long-Baseline-Interferometry. Astron. J. 76-2, p. 91–100.

(310) Campbell, J.: 1979. Die Radiointerferometrie auf langen Basen als geodätisches Messprinzip hoher Genauigkeit. Munich/F.R.G., Verl. Bay. Akad. d. Wiss., Reihe C, Ha Nr. 254, habil. thesis, Landw. Fak., Univ. Bonn (in German), 120 p.

(311) Schuh, H.: 1987. Die Radiointerferometrie auf langen Basen zur Bestimmung von Punktverschiebungen und Erdrotationsparametern. Diss. thesis, Landw. Fak., Univ. Bonn/F.R.G., Verl. Bergk/Munich, ISBN 37696 93787, 124 p.

(312) Zensus, J.A., Pearson, T.J. (Eds.): 1987. Superluminal Radio Sources. Proc. Workshop in honour of M.C. Cohen, Big Bear Solar Obs. Cal./USA, Oct., Cambridge Univ. Press/Mass., 361 p.

(313) Preuss, E.: 1988. VLBI from ground and space. ESA, Space Science and Fundamental Physics. Proc. Summerschool Alpach/Austria, Aug., 12 p.

(314) Schnizer, B., Pascher, W., Wohlleben, R.: 1986. Theoretical and Experimental Investigation of a Torus as a Primary Feed in Reflector Antennas. Proc. URSI Symp. Electromagnetic Wave Theory, Pt. B., Budapest, Aug. 28, p. 707–710.

(315) Wohlleben, R., Klein, U., Neidhöfer, J.: 1988. Multiple-Beam and Polarization Characteristics of the 100-m Effelsberg Radiotelescope. Proc. IEEE AP-S, Internat. Symp., Vol. 3, Suracuse/N.Y.-USA, p. 1100–1102.

(316) Morris, D., Baars, J.W.M., Hein, H., Steppe, H., Thum, C., Wohlleben, R.: 1988. Radio-holographic reflector measurement of the 30-m millimeter radio telescope at 22. GHz with a cosmic signal source. Astron. Astrophys. 203, p. 399–406.

(317) Reid, M.J., Moran, J.M.: 1987. The impact of VLBI on astrophysics and geodesy. IAU Symp. 129, G.P.O. Washington DC, 600 p.

(318) Levy, G.S., Linfield, R.P., Edwards, C.D., Ulvestad, J.S., Jordan, J.R., Dinardo, S.J., Christensen, C.S., Preston, R.A., Skjerve, L.J., Stavert, L.R., Burke, B.F., Whitney, A.R., Cappalla, R.P., Rogers, A.E.E., Blaney, K.B., Maher, M.J., Ottenhoff, C.H., Jauncey, D.L., Peters, W.L., Reynolds, J., Nishimura, T., Hayashi, T., Takano, T., Yamada, T., Hirabayashi, H., Morimoto, M., Tokumaru, M., Takahashi, F.: 1989. VLBI in earth orbit. I. Astrophys. Journ. 336, p. 1098–1104.

(319) Linfield, R.P., Levy, G.S., Ulvestad, J.S., Edwards, C.D., Dinardo, S.J., Stavert, L.R., Ottenhoff, C.H., Whitney, A.R., Cappallo, R.J., Rogers, A.E.E., Hirabayashi, H., Morimoto, M., Inoue, M., Jauncey, D.L., Nishimura, T.: 1989. VLBI using a telescope in orbit. II. Astrophys. Journ. 336, p. 1105–1112.

(320) Bielli, P., Massaglia, M., Savini, D., Klooster, K.V.T.: 1986. Coaxial horns with corrugated Flange. Proc. of URSI Int. Symp. on Electromagnetic Theory, (Ed. Berceli, T.), Budapest, Elsevier/Amsterdam, p. 718–720.

(321) Alef, W., Porcas, R.W.: 1986. VLBI fringe-fitting with antenna-based residuals, Astron. Astrophys. 168, p. 365–368.

(322) Alef, W., Götz, M.M.A., Preuss, E., Kellermann, K.I.: 1988. Structural changes in the nucleus of the double radio galaxy 3C390.3. Astron. Astrophys. 192, p. 53–56.

(323) Witzel, A., Schalinski, C.J., Johnston, K.J., Biermann, P.L., Krichbaum, T.P., Hummel, C.A., Eckart, A.: 1988. Astron. Astrophys. 206, p. 245–252.

Index

Abberation, relativistic 111
absolute frequency phase 59
aliasing 99
allpass 165, 166
ambiguity 135, 142, 158, 159, 167, 185
    function (af) 8, 136, 148, 151, 152, 160, 161, 162, 168
    of direction finding (DF) 135, 182, 184
    , phase 130, 136
amplifier 48, 101, 175
amplitude calibration 89, 91
    spectrum 101, 144, 158
analytical signal 149
angle, hour 76, 112
    of incidence 11, 25, 34, 131, 137, 165, 169
    , parallactic 41
    , solid 11, 16, 176
    , target 161
angular ambiguity 135, 139, 186, 187
    diameter 41, 66
    extension 32, 41
    frequency 119, 179
    momentum vector 110
    resolution 8, 17, 24, 56, 58, 115, 136, 169
    velocity of earth 26, 37, 38, 72, 115
    width 41
antenna axis, offset 27
aperture 8, 9, 15, 167, 169, 171, 175, 178, 183
    , circular 169, 171, 177
    , effective 9, 16
    , filled 9, 39, 43
    illumination 157, 159, 161, 165
    , single 157
    synthesis (AS) 41, 56, 75, 85, 88, 91, 116
    mapping 90
    MERLIN network/G.B. 85, 89
    VLA/New Mexico-USA 89
apogee 115, 116
a priori model 101, 110, 111
array, antenna 22, 24, 35, 39, 40, 44, 89, 90, 153, 158, 159, 160, 165
    factor 25, 40, 137
ascension, right 44, 102, 112
astrometric experiment 106, 114
atmosphere, earth 88, 114, 143, 169, 171
    , turbulent 51, 114, 171
atmospheric errors 34, 51, 88, 104, 136, 143, 169
    noise 114

atomic frequency standard 59, 84, 110
    time 111
autocorrelation 20, 44, 87, 102, 146, 156
    function (acf) 20, 44, 87, 136, 146, 152, 157, 159, 179, 181
    tack 153, 154, 164
automatic gain control (AGC)
average, time 46, 62, 63
axis roll 184
    , rotation of the earth 61
    , yaw 184
azimuth 175, 177
Band 19
    , double side (DSB) 79, 82
    , lower side (LSB) 79, 119
    , single side- (SSB) 79, 82
    , upper side- (USB) 79, 119
bandpass envelope 19, 81, 101, 122
    width 18, 28, 51, 53, 68, 119, 122, 154, 160, 168
    , signal-to-product 155, 160
Barker code 155, 156
baseline 8, 9, 25, 34, 51, 96, 130, 138, 179, 185
    error 51, 135
    gain factor 90
    , intercontinental 57, 96
    , projected 18, 59, 62
    , retarded 114
    vector 25, 61, 71, 106, 113, 127
Bayesian statistics 101
beam 29, 96, 135, 136
    , Gaussian 18, 97
    , true 95, 100
bearing 177, 178, 182
binary phase switching 155, 161
Boltzmann's constant 12
Bose-Einstein statistics 101
Bracewell's principal solution 18, 43, 96, 98
brightness distribution 11, 16, 18, 30, 52, 63, 70, 93, 101
    distribution temperature 11, 16, 24
    temperature distribution 11
    of a source 30, 101
    of sky 19, 70, 91, 99, 101
Cable delay 34, 54, 110
caesium beam clock 87
calibrated visibility 88
calibration 59, 88, 89
    , redundant 89
    source 89
    , self- 59
Cambridge 10, 35

carrier frequency 148, 177, 179, 181
cassettes (for TV) 88
Cassiopeia 35
celestial pole 70
cell size 99
center, phase 71, 75, 143
central lobe (of dirty beam) 100, 140
characteristics, power 16
Christiansen cross interferometer 41
circular polarization 17, 42, 59
Cittert-Zernike theorem 18, 65
clean 59, 91, 93, 94, 97, 99, 100
    beam 99
    window 100
clipper 87
clipping 87
clock 84, 87
    error 84, 90, 104, 105
    offset 88
    synchronization 84, 87, 113
closure amplitudes 90
    phases 59, 82, 89, 90
code, Barker 155, 156
    , pseudonoise 136, 177, 179
coherence 10, 14, 18, 63, 117, 122, 180
    function of a source 18, 63
    function, lateral (mutual) 18, 64, 88
    length 10
    , partially 63
    , temporal 68
    theory 18, 62, 68
    time 10, 84
compact source 33, 89, 115
    structures 33, 115
completely incoherent field 64
complex correlation function 23, 102
    visibility function 75, 80, 88
    visibility, observed 80
compound interferometer 53
confusion probability 148, 153
conversion 119, 121
convolution 19, 30, 96, 146
    equation 96
    function 98, 99
    theorem 20, 149, 159
co-operative system 184
coordinated, universal time 61
coordinates, equatorial 70, 110
    , spherical 13, 18, 59, 115, 137
    , transform matrix 71
correction phase 135
correlation 23, 44, 88, 104, 122
    , auto 20, 44, 87, 102
    center 83
    equation 146

, simultaneous 165
correlated data streams 83, 88
    spectrum 49, 101
correlator 47, 61, 88, 104, 114, 119,
    127
    , digital 88, 104
    , Mark II 70, 87
    , Mark III 76, 88
crest of wave 114
crosscorrelation 23, 44, 50, 62, 66, 82,
    88, 101, 121, 136, 146, 163
    coefficient (CCC) 44
    spectrum 67, 104
    interferometer 41, 61, 66, 67, 104
crustal motion 110
crystal oscillator 84
Data digitizer 84, 88
    processing 83, 117
    recording 87, 116
day, sideral 113
declination 38, 44, 76, 102, 113
deconvolution 93, 99
degree of coherence 63
delay 68, 119, 127, 160, 161
    , a priori model 111
    atmospheric phase 104, 105
    beam effect 127
    error 80, 105
    , geometrical 61, 68, 80, 88, 104,
        112, 119, 122
    group 59, 106
    propagation 111
    range
    residual 113
    , system 68, 123
    time 61, 160
    tracking error 104, 105, 123, 124
    tracking error, perfect 123
delta function, Diracs 19, 65, 96, 99
depolarization factor 17
difference, path 26, 36, 73, 169, 171
differential fringe rate 105
diffraction 9, 18
    grating 41
    limit 56
    correlator 104
digitize 87, 88
dipole 35, 42
Dirac delta function 19, 65, 96, 99
direction finding 182
directivity 13, 43
dirty map 93, 99, 100
discovering probability 143
distance resolution 8, 152, 155

distribution, brightness temperature 11,
    16, 24
diurnal motion 122
diversity, polarization 188
Doppler effect 88, 115, 131, 148, 150,
    161
    resolution 148
    shift 88, 105, 131, 136
double sideband receiver (DSB) 79, 119,
    123
duplexer 175
Earth angular velocity 26, 37, 115
    plates, tectonic 110
    rotation 25, 26, 68, 88, 104, 106,
        114
    tides 110
east-west interferometer 25, 38, 41, 44,
    76, 127
echo signal 146
ecliptic plane 24
effective aperture 15
    height (of an antenna) 13, 26
Effelsberg viii, 56, 103
Einstein statistics, Bose- 101
electrical interference 27, 29, 84
elevation 34, 169, 171, 173, 177, 180,
    181
ellipse of projected baselines 75
    of $u - v$ plane 59, 75, 93
elliptical track 75, 77
ellipticity 13
entropy 100, 101
    map, maximum 100
envelope, modulation 42
equatorial component 70, 76
    coordinates 70
equivalent circuit 26
extended source 16, 30, 32, 43, 71, 101
European VLBI network (EVN) 57
extraterrestrial media 88
False echo 143, 146, 148, 150
Faraday rotation 187
farfield 18, 20, 66, 158, 178
feeds 116, 177
field of view, primary 96, 99, 127
filter of higher harmonics 120
    comb 167
    matched 143, 181
    passband (IF) 51, 101
    , pulse compression 154
    RF 22, 67, 143
    , spatial 18, 20, 62
final map 100
FK4 system 61
flux density 11, 12

formatter 87
Fourier inversion 19, 47, 73
    pair 63, 64, 91, 104
    spectrum 20, 31
    transformation (FT) 18, 19, 45, 63,
        79, 91, 98, 122
    , inverse 19, 47, 73, 91
    theory 78, 98, 157
frames, block 88
free space impedance 14
frequency 19, 23, 67
    , fringe 114
    , heterodyne 79, 119
    , intermediate 50, 79, 84, 120, 125,
        129, 175
    modulation 155
    , radio (RF) 28, 53, 118, 125
    spatial 19, 20, 39, 63
    standards 84, 105
    , atomic 84
fringe amplitude 26, 50, 66, 82, 122,
    124
    , differential 105
    , diffraction 9
    , distance 36
    frequency 114, 119, 125, 126
    pattern interferences 10, 59, 122
    phase 59, 81
    rate 37, 59, 62, 102, 105, 111
    pattern, interference 62
    , geometrical (natural) 62, 66
    method 105
    , natural 62, 81, 88, 119
    rate 72, 88, 101, 113, 119, 126
    search 84
    spatial 63
    washing function 68, 80, 101
    white 70
fringes 9, 38, 51, 62, 122
function, correlation 23, 102
    crosscorrelation 23, 88, 101, 103
    , visibility 88, 91, 102
    , complex 102
Gain 176
    , loop 100
Gaussian beam 97
    function, tapered 69, 99, 141
Gemini satellite 10
geodesy 106
geodetic error 59, 106, 182, 183
geometrical
    time delay 61, 68, 80, 88, 104, 112,
        122
    optics 187
    path difference 73, 104

time delay
grating interferometer 38
    lobe 40, 135, 137, 148, 158, 161,
        163, 163, 167
grid, equally spaced 91, 98
gridding 98, 99
group delay 59, 106
H-MASER clock 59, 84
half-power beam width (HPBW) 8, 28,
        40, 135, 148, 159, 160, 165
halo 33
Hanbury-Brown-Twiss interferometer
        51
harmonics, higher 120
Hertz, milli- (mHz) 112
heterodyne frequency 79
    technique 79, 119
horn aperture 116, 177
    , feed four 177
    , hybrid (HE$_{11}$) 116
hour angle 76, 112
hybrid 91
    mapping 91
    wave horn (HE$_{11}$) 116
Image 18, 19, 24, 43, 79, 89, 93, 96,
        98, 116
impedance, free space 14
impulse compression filter 154
incidence, angle of 25, 34, 61
incoherent source radiation 16, 19, 23
index of refraction 34, 169, 170
integrated system 130, 177
integration time 50, 51, 84, 126
integrator 48, 67
intensity fluctuation 50
    interferometer 50
    of source 18
intercontinental baseline(s) 57
interference amplitude 27, 29, 84
    , constructive 62
    destructive factor 62
    fringe patterns 9, 38, 64, 122
    lobe 122
    modulation factor 84
    , order of 26, 29
interferometer 9, 10, 21, 25, 35, 136
    , active (or radar) 130, 178
    , broadband 36, 165
    , Christiansen cross 41
    , compound 53
    , connected element 34
    , correlation 47, 62, 67
    , cross 41
    , double baseline 29, 41, 136
    , east-west 25, 38, 41

equation 9, 62, 136, 184, 186
    , frequency-swept 36
fringe 26, 51, 62, 105
geometry 21, 59, 70
grating 38, 40, 115, 135
    , Hanbury-Brown-Twiss 51
hour angle 76
    , intensity 50
    , linear 160, 161, 166
    , link 34, 51, 53
    , Lloyd's mirror 34
lobes 25, 26, 28
    , long baseline (LBI) 53, 83, 186
    , medium baseline (MBI) 85
    , meridian transit 28, 38
    , Michelson's stellar 9, 28, 31, 56
    , Mill's cross 41
    , multi-element (or grating) 38, 116
    , phase coherent 118
    , phase-switched 35
    , pinhole 63
    , post detection correlation 50
    , principle of 9, 25
    , pseudo-noise code 177
    , radar
    , radio 24, 34
    redundant 89
    response 72, 84
    resulting diameter 43, 59
    , Ryle's 28, 35, 56
    , satellite 188
    , sea 33
    , short baseline (SBI) 136, 186
    , stellar-swept frequency 36
    , switched 22, 35
    , three-element (or triple) 39, 116,
        139
    , twin 10, 25, 26, 35, 56, 118, 132,
        137, 150, 169, 177, 178
    types 21, 28
    , Very Long Baseline (VLBI) 53, 56,
        83
    with lobe switching 28, 35
intermediate frequency (IF) 50, 79, 84,
        120, 125, 129, 175
ionosphere 10, 110, 187
iteration, Number of, N 92, 100
inverse Fourier transform 19, 47, 96
Jansky (unit) 12, 103
Jet physics
Khinchin relation, Wiener- 101
Lateral coherence function 18, 64
least squares technique 89, 90, 113
light deflection, relativistic 111
line VLBI, spectral 101, 103, 115

link interferometer 34, 51, 53, 117
Lloyd interferometer 34
lobe, interference 26, 96
    unique 100, 159
    switched interferometer 35, 38
local oscillator (LO) 66, 79, 88, 90, 98,
        104, 105, 175
    frequency 66, 104
    stability 98
loop gain 100
LORAN-C (time) 87
low pass 22, 39
lower sideband (receiver) LSB) 79, 119
    response 80
Magnetic tape 59, 84, 88
magnetron 176
main azimuth
    , beam 19, 96, 153
    , lobe 139, 148
map, clean 89
    delta 100
    , dirty 93, 99, 100
    , hybrid 91
    of source brightness 88, 115
    , positivity 101
    , residual 95, 99
    , superresolved 100
mapper 100
mapping 88, 89, 90, 101, 115
Mark II VLBI correlator 70, 82, 87,
        105
Mark III VLBI correlator 70, 82, 88
MASER clock (H-) 59, 84
    , OH- 64, 101, 103, 106
    , Orion 64, 101, 103
    , sky 64, 101, 103, 106
master station 87
matched filter 143, 144, 148, 151, 152,
        154, 158, 161, 163, 165, 169,
        181
    filter bank 151, 161
    receiver 143, 148, 184
matrix coordinate transform 71
maximum entropy map, MEM 59, 91,
        100
medium baseline interferometer (MBI)
        85
meridian plane 25, 43
    transit interferometer 41
MERLIN network/GB (aperture syn-
        thesis) 85, 89
Michelson stellar interferometer 9, 53,
        56
milli arc second (MAS) 59, 89, 105,
        107, 115

Hertz (mHz) 112
Mill's cross interferometer 41
minimum 28
Minitrack network 10, 131
mismatch, polarisation
mixer 79, 125, 175
mixing 79, 175
model, a priori 89, 110
    delay 110
    , final 92
    fitting 74
    , starting 89
modulation 29, 42, 122, 136, 148, 155,
        158, 168, 179
modulator 35, 184
moment, 4th order 52
monochromatic wave 10, 13, 28, 61
    , quasi- wave 13, 18
    , source 13
monopulse radar 130, 177
    , one-dimensional 130
    , simultaneous 130
    , two-dimensional 175, 176, 177
motion, polar 61, 106
    diurnal 122
    of atmosphere 51, 114
    , radial 105
    term 59
multibeam 39, 116
multichannel receiver 151, 161
multi-element interferometer (grating
        i.) 38
multilobing 25, 44, 136, 138, 159
multiplier 47, 119
multiplicative law 25, 43, 136
mutual coherence function (of a source)
        65, 67
Natural fringe rate 62, 81, 119
navigation 181, 187
navigational system 181, 184, 187
network, MERLIN/GB 85, 89
noise 35, 90, 99, 100, 114, 136, 144
    currents 48
    , input power 35, 89
    , output power 146
    , phase 84
    , receiver 35, 48, 89, 96, 122
    , temperature 16
    , voltage 100
normal distribution 141
nutation 106
    of earth axis 61
Nyquist rate 87
    sampling theorem 23, 56, 87, 181
Observed complex visibility 88, 89

offset, clock 88
    of a local oscillator 88, 90, 105
    of the antenna axis 140, 160, 176
OH-MASER 64
one-bit correlation 87
    representation 87
optical telescope 9, 10
order of interference 26, 29
Orion MASER 64, 106
orthogonal polarization
oscillator (, local) (LO) 66, 79, 88, 98,
        104, 119, 175
Pair, Fourier 63, 104
parallactical angle 41
Parceval's theorem 147, 149
passband, filter 19, 101
path
    difference 10, 26, 73
    delay 114, 169
    length 73, 114, 131, 134, 169
    difference, geometrical 73
    time difference 114
pattern, antenna 16, 20, 157
    , array 25, 40, 137
    , interferometer 21
pencil beam 18, 19, 43, 44, 132
perfect delay tracking 82
phase, absolute 59, 89, 104, 105
    ambiguity 130
    calibration 59, 89, 71, 143
    center 71, 143
    closure 59, 89
    comparison 130, 175, 176, 184
    difference 25, 184, 185
    jumps 82, 123, 125
    locked loop (PLL) 84
    matching 125
    modulation 156
    noise 84
    referencing 59, 82, 93, 104
    reference point 70, 93, 104
    rotator 119, 125, 126
    selective detector (PSD) 43
    self-calibration 59, 82, 89, 91
    spectrum 144
    stability 51, 104, 126
    sum 90, 135
    switching, binary 35, 155, 161
phases, station 90
phasing 125
pinhole interferometer 63
pixel(s) 98, 100
Planck radiation law 11
plane of source 18
play back system 84

point-like source 89, 105, 178
point source 59, 89, 102
    component 102
    model 89, 110
    spread function 96
pointing accuracy 8, 61, 118, 185
    error 132
polar motion 61, 106, 113
polarization, degree of 15
    ellipse 12, 13
    mismatch 15
    ratio 15
    vector 42, 61, 63
polarized wave, partly 13, 14, 17
    , statistically 12, 16, 18
positivity 101
power characteristics 14, 16, 20
precession 61, 106
predicted visibility 101
principle solution, Bracewells 96
primary standard 59, 84, 85
probability, a priori 141
    of ambiguities 141
    confusion 148, 153
    discovery 146
projected baseline 18, 59, 62
pseudo-noise code 136, 156, 177, 178,
        179
Pulkovo observatory 34, 40
pulsar 115
pulse compression 154
    filter 155
    response 132, 144, 153, 160, 163
Quantum jump 10
quasar 10, 58, 59, 78, 95, 115
quasat (satellite) 116
quasi-monochromatic wave 13
RADAR 130, 153, 159, 172, 175
    equation 175
    , monopulse 130, 175, 176
    , monopulse, one-dimensional 130
    , two-dimensional 175, 176
    range 130, 159, 171, 174, 175, 176
radial motion 103, 105
    of source 103, 105
    velocity 103
radiation density 12
    resistance 13, 26
radio amplifier 48
    frequency (RF) 48, 101, 175
    source 11, 67, 101
    telescope 8
range radar 130, 159, 171, 174, 176,
        179
rate, natural fringe 62, 72, 105

Nyquist 87

ratio, signal-noise (SNR) 50, 87, 98, 123, 126, 143, 146, 154

ray 25, 132, 135, 169

Rayleigh criterium (far field) 8, 18, 20, 66, 158

Rayleigh-Jeans law 11

ray matching 169

real-time (datalink) 34, 51, 59

receiver, correlator 18, 88, 101, 116
, crosscorrelator 83, 119
, DSB 79, 82
, SSB 79, 82
, optimal 143
, RADAR 143, 148

reciprocity theorem 13, 167

reference frequency 84, 105

referencing, phase 59, 70, 105

refraction, atmospheric 34, 131, 136, 169, 170, 173
error 131, 136, 169, 173
index, averaged of 171

relativistic abberation 111
effect 111
light deflection 111

resampling 56

residual map 99

resolution, angular 8, 9, 10, 76, 115, 116, 130, 135
, distance 8, 152, 155, 181
, general 8
, spectral 8, 102
, velocity 8

response impulse 132
, signal 28, 43, 66, 70

resulting diameter of an interferometer 43, 59

RF signal 28, 53, 118, 125

right ascension (RA) 44, 102, 103, 112

roll axis 184

rotation axis of earth 106

rubidium clock 84
standard 84

Ryle 28

Sampling frequency 23, 56. 181
function (ordinary) 91, 96
weighted 43, 96

satellite radar 180, 188

Schwarz' inequality 44, 147

sea interferometer, Lloyds 33

self-calibration 89, 90, 91

sensitivity 40, 102, 111

servoloop 76

shi function, Bracewells 98

shifting theorem 18, 20

short baseline interferometer 186
waves 29

sideband 125
, lower (LSB) 119
separation 119, 125
upper (USB) 119

sidelobes 43, 96, 98, 103, 135, 140, 158, 160, 175

sideral day 113

si-function 23, 29, 99

signal, form 22, 101, 136, 148, 158
processing 88, 118, 161
, sum 177
theory 144, 149

signal-to-noise ratio (SNR) 50, 87, 98, 123, 126, 143, 146, 154

sine function 23, 29

single pulse 144, 146
sideband receiver (SSB) 79, 82, 122, 127
mixer 79

size of a cell 99

sky MASER

slave station 87

smoothness 98, 101

solar observation 34, 37, 41, 43

solid angle 11, 16

source, angular size of 12, 111
, brightness distribution of 18, 63, 65, 91
map 100
coherence function (mutual) 63
, tracking 10, 28
constancy
, coherent 88
, discrete 12
, point 28, 30, 59, 89, 102
, extended 12, 30, 41, 72
, incoherent 16
model 33, 100
, monochromatic 13, 28
, plane 61
, radio 11, 33, 115
structures, compact 33, 59, 89
vector 70

spaced grid, equally 90, 91, 163

space VLBI 115

spatial characteristics 41, 42, 43
coherence 88
frequency filter theory 18, 19, 20, 21, 31, 62, 63
period 18, 19
spectrum 19, 21, 40, 55

spectral line VLBI 59, 67, 101, 103, 115

resolution 28, 159
spectroscopic VLBI 59, 101, 115
spectrum, amplitude 11, 47, 101, 144, 159
, energy density 46
, crosscorrelated 49, 101
, rectangular 159
, spatial 21
spherical coordinates 19
spreading 96
spread, point function 96
square fit, least 89, 90, 113
stability of local oscillator 84, 98
standard frequency 84, 105
atomic 84, 105
starting model 89
station, master 87
stationary 63, 161
stations, VLBI 60, 88, 90, 108
statistical independent noise 35, 90, 99
polarization 12, 16, 18
statistics, Bayesian 101
, Bose-Einstein 101
superluminal motion 59
superresolved map 100
supersynthesis 75
surface acoustic waves (SAW) 155
sweep interferometer 36, 38
swept frequency interferometer 36
synchronous satellite 181, 186, 187
synchronization 84, 88
, clock 84, 113
synthesis, aperture 41, 56, 88, 116
mapping 88, 116
radio telescope 56
Westerbork (WSRT) viii, 89
VLA/USA 89
super 75
synthesized half-power beam width 59
synthesized instrument 41
system, FK-4 61
, VLBI Mark II 70, 87
, VLBI Mark III 70, 88
Tack, ambiguity 153, 154, 164
tape, magnetic 59, 83, 88
tapered Gaussian sinc function 98, 99
target 130, 131, 132, 135, 148, 152, 161, 162, 164, 165, 169, 171, 174
technique, heterodyne 79
tectonic earth plates 110
telescope, optical 9, 96
, radio 9, 59
temperature, antenna 16
, noise 16

, radiation 87
temporal average 46, 62, 63
coherence condition 10
theorem, Cittert-Zernike
, convolution 20, 149
, sampling 23, 56
theory, coherence 18, 63
, Fourier 19, 78
thermal noise 16, 87
three-element interferometer 39, 116, 139
threshold value 143
tides 109
time delay 61, 68, 132, 144
, real 59
synchronization 84, 88
, universal coordinated (UTC) 61
time-bandwidth-product 155, 160, 168, 180
topological uncertainty 141
tracking 10, 28, 61, 115, 130, 132, 176, 177
, source 28, 66, 84, 122, 176
, perfect delay 104
tracks (ellipses) 75, 77
, audio 88
, $u - v$- 77, 88
transfer function 93, 96, 144, 146, 165, 166, 167, 168, 169
transform, fast Fourier (FFT) 98
, Fourier (FT) 18, 19, 45, 63, 79, 96, 98, 103, 122, 145
, inverse Fourier 19, 47, 73
matrix, coordinates 71
transponder 130, 177, 179, 188
tropospheric delay 10, 110
true beam 89
turbulent atmosphere 51, 169
twin interferometer 10, 28, 35, 56, 160, 169, 178
two-pi ambiguity 135
Uniform weighting 98
universal time (UTC) 61
upper sideband (USB) 79, 119
UT1 time 110
UTC time 61
UTR 1, 2, 29
$u - v$-ellipse 59, 75, 93
plane 59, 75, 77, 88, 89, 98, 115
plot 59
spacing 78, 100
track 75, 79, 96
Van Cittert-Zernike theorem 18, 65
van Vleck equation 87
theorem 87

variance 141
   of visibility 141
vector, baseline 61, 71, 113
   , source 70, 106
velocity of light 12, 62, 131, 165
vertical pattern 136
very large array (VLA) 89
very long baseline array (VLBA) 115
   interferometry (VLBI) 57
   interferometry receiver 88
view, primary field of 96, 99, 127
visibility 18, 31, 64, 78, 88
   , calibrated 88
   function 32, 59, 73, 88, 90
   , complex 75, 80, 88, 90
   , weighted 90, 98
   , observed 88, 89
   , predicted 101
   , true 89
   , variance of 141
VLBI experiment 56, 79, 83, 89, 106,
      113, 116
   correlator 88, 104
   Mark II system 82, 87, 105
   Mark III system 82, 88
   , space 115
   , spectral line 59, 67, 101
Washing function 68, 80, 101
wave crest 114
wave packet 133
weighting function 96
   sampling functions 96
   visibility functions 98
west interferometer, east- 25, 38, 44,
      76
Westerbork viii, 89
Westerbork (aperture) synthesis radio
      telescope (WSRT) viii, 89
white fringes 70, 122, 124
width, band- 18, 28, 101
Wiener-Khinchin-relation 101
Wiener's theorem 45, 47
Wilson, Mt., observatory 9, 53
Yaw axis 184
Zenith 10, 43, 135, 160, 161, 162
Zenith angle 43
Zernike-theorem, Cittert- 18, 65